Data Driven Energy Centered Maintenance

2nd Edition of Energy Centered Maintenance: A Green Maintenance System

Data Driven Energy Centered Maintenance

2nd Edition of Energy Centered Maintenance: A Green Maintenance System

Fadi Alshakhshir
Emaar Properties PJSC, Dubai,
United Arab Emirates
and
Marvin T. Howell
Energy and solar consultant, USA

River Publishers
Routledge
Taylor & Francis Group
LONDON AND NEW YORK

Published 2021 by River Publishers
River Publishers
Alsbjergvej 10, 9260 Gistrup, Denmark
www.riverpublishers.com

Distributed exclusively by Routledge
4 Park Square, Milton Park, Abingdon, Oxon OX14 4RN
605 Third Avenue, New York, NY 10017, USA

Library of Congress Cataloging-in-Publication Data

Data Driven Energy Centered Maintenance / by Fadi Alshakhshir and Marvin T. Howell --- 2nd edition

Routledge is an imprint of the Taylor & Francis Group, an informa business

ISBN 978-8-7702-2357-7 (print)
ISBN 978-8-7702-2356-0 (online)
ISBN 978-1-0031-9510-8 (ebook master)

Table of Contents

Dedication

Dedication from Fadi Al Shakhshir

This book is dedicated to the memory of **my father**, to **my mother**, my family, my colleagues, and my friends for giving me the strength to complete this work.

I also would like to dedicate this book to **Mr. Ahmad Juma Al Falasi** for his motivation and support.

Dedication from Marvin Howell

This book is dedicated to Valerie Oviatt, Director, Seminars and Internet Training, AEE Energy Training who selected me as an AEE Energy Instructor and approved Energy Centered Maintenance (ECM) to be an Online Energy Seminar.

In addition, I would like to thank my wife, Louise Howell, for her continued support.

Preface

Energy Centered Maintenance (ECM) originated in 2012 by Marvin Howell. He has a BS degree in mechanical engineering from Mississippi State University with an MS degree in industrial engineering from University of Pittsburgh. He has over 30 years' experience in maintenance management and facilities maintenance including over 14 years in energy and environmental management. Marvin Howell kept finding motors running 24/7 when they were only required to run for 7–8 hrs daily. Also, he observed switches stuck on equipment, sensors not working, building automation systems with operators not trained, data centers using servers that were energy hogs, and cold air mixing wrongly with the hot air on the way to the computer room air conditioner (CRAC). He recognized that a maintenance program is needed to address equipment using excessive energy. Of the six different maintenance systems, none addressed this energy waste as the primary focus. Marvin was excited to present this concept on Association of Energy Engineers (AEE) On-Line Energy Seminars.

Fadi Al Shakhshir, holder of a bachelor's degree in mechanical engineering from Jordan University of Science and Technology, an M.Sc. degree in energy from Herriot Watt University, and a Certified Energy Manager from AEE, attended in early 2016 the ECM Seminar that Marv instructed. By the end of 2015, Fadi was already developing a maintenance program that focuses on including energy-related maintenance tasks to regular reliability maintenance plans and called it energy centered maintenance which turned out to be a similar program to what Marv invented. The initial ECM concept was enhanced tremendously by his development of the technical steps necessary to implement ECM and the extension of its application to additional equipment and extending the idea outside of building systems to water supply system, drainage system, and the fire protection system. It was apparent that the world could benefit if both Marv and Fadi would come together and write a comprehensive ECM book outlining how to implement and sustain this new and beneficial green energy maintenance program.

In 2020, Fadi Alshakhshir defined a new concept that converts energy centered maintenance model from a manual system, into a data driven model; accordingly, the revised edition of the book has the title "Data Driven Energy Centered Maintenance."

Glossary

Annual Work Plan	The proactive maintenance schedule on a 52-week calendar. It lists every facility, system, and equipment with its job plan and frequency for each proactive maintenance activity to be performed.
Asset	Any facility, system, equipment, or component.
CMMS	Computerized maintenance management system.
Corrective Action	Repairs made when an asset fails to operate as intended.
Craftsperson	Any qualified technician assigned to handle problem calls and PM procedures.
Equipment	The individual components of mechanical and electrical systems that are serving the building to function. Examples are heating, ventilating, and air conditioning (HVAC) systems; elevators; and communications systems.
Facility	The buildings, utilities, structures, and other land improvements associated with a building, operation, or service.
Failure	Event rendering equipment non-useful for its intended or specified purpose during a designated time interval.
Flowmeter	A device used to measure the flow rate of a liquid or air.
Functional Failure	The loss of function as the inability of an asset to meet a desired standard performance.
Job Plan	It provides all the details regarding operations, materials, labor, and tools required to do the work.
Maintenance Management	The administration of a program using such concepts as an organization, plans, procedures, schedules, cost control, periodic evaluation, and feedback for the effective performance and control of maintenance with adequate provisions for interface with related disciplines such as health, safety, environmental compliance, quality control, and security.

Maintenance Supervisor	The individual having authority and responsibility for specific maintenance activities at a facility or system.
Predictive Maintenance	Condition-based maintenance strategy where one or more equipment outputs is measured about the degradation of a component or subsystem.
Preventive Maintenance	Time, usage, or cycle-based maintenance strategy in which periodic testing, servicing, adjustments, lubrication, or inspections are performed on equipment to determine the progress of wear in components or subsystems. It can prevent/mitigate failure or detect hidden failures.
Planning	The identification and assessment of needed resources and the order in which the resources are required to complete a job plan in the most efficient manner. Planning defines the scope of work on a work order, the resources needed to complete the job, the sequence of the jobs, how long each job will take, and which jobs can be done concurrently.
Potential Failure	Point at which the facility, system, or equipment has been detected as failing.
Productive Work	Work that corresponds to a work order or is related to a piece of equipment.
Reactive Maintenance	Maintenance strategy where equipment is allowed to operate with no associated maintenance program. Also referred to as "Run-to-Fail."
Root-Cause Analysis	The methodology used to identify solutions to prevent failure from occurring. It is not root causes that are sought: it is effective, controllable, goal meeting solutions to prevent failures. RCA process is also used in RCM model to prevent operational deficiencies from occurring and to eliminate or reduce energy waste.
Reliability-Centered Maintenance (RCM)	A structured/logic-based process used to develop complete system and equipment maintenance programs providing the highest level of equipment reliability at best possible cost.
Scheduling	The assignment of job plans to a specific period of time to maximize the use of available resources.

Scheduled Work	Work that can be identified, predicted, or planned well in advance.
Scheduling	The assignment of definite amounts of work to personnel based on estimates of how many personnel labor hours are available for the planning horizon.
System	A logical and systematic group of assets that are necessary to support the facility's mission. A system must be described in a breakdown structure for each site so that it can be properly identified and managed.
Site	All structures and systems that support a building or a facility within a facility, e.g., utilities, parking lots, roadways, bridges, fences, tunnel, etc.
Tools	Inventoried implements used to perform or assist in performing maintenance work functions within the facility, e.g., specialized hand tools, calibration tools, power tools, electric cords, mechanical tools, etc.
Tasks	Instructions to be followed in the performance of maintenance procedures.
Wrench Time	Productive work. This work is the actual mechanical work performed by a technician, manager, or contractor.

List of Tables

List of Figures

1

Energy Reduction

1.1 Energy Cost

In the United States, around $500 billion a year is spent on energy. Energy costs normally represent up to 30% of most corporations' operating expenses. The U.S. Green Buildings Council estimates that commercial office buildings use, on the average, over 20% more energy than they should, which is a significant dollar loss to industry due, primarily, to the fact that management does not know where the waste is occurring and how to eliminate or reduce this loss.

The three ways to reduce energy consumption are shown in Figure 1.1.

A common **Goal** established by many organizations is to reduce energy consumption by 10% in the next two years compared to the energy baseline of last year. The goal drives action and the components contributing look like the following: Goal---Strategy---Objectives---Action Plans or Projects.

A **Strategy** is simply a plan of how you are going to achieve the goal. Our present strategies shown in Figure 1.1 are as follows:

1. Implement the low hanging fruit and address identified energy waste.
2. Implement energy conservation program in our organization.
3. Implement energy efficiency measures which include energy efficiency projects and energy centered maintenance.

1.2 Implementing Low Hanging Fruit

Low hanging fruits are the most cost-effective actions that can reduce energy use and costs. These measures can be applied immediately such as behavioral changes as switching off unnecessary lights or adjusting set points and time schedules for HVAC systems or that require little investment such as conducting testing and balancing for some mechanical

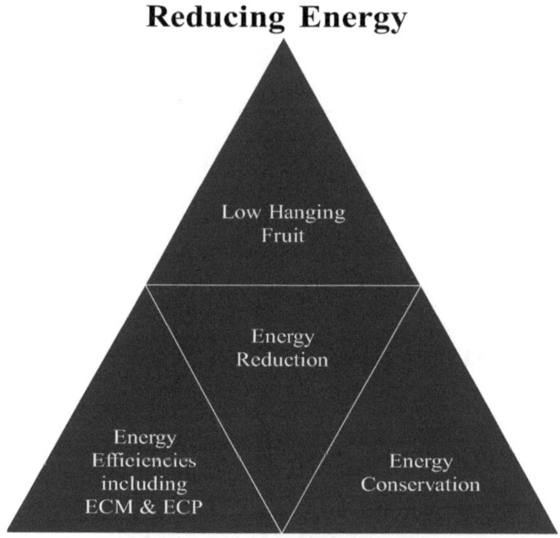

Figure 1.1 Energy reductions.

systems. The simple payback can be immediate and seen in the future energy bill.

Low hanging fruits are those hundreds of things that are available for an organization to select (pick) and implement at no or little cost but do reduce energy. They are:

- Not already implemented.
- Easy to implement.
- For my organization, low cost or no cost.
- Can sell to my management.
- Will reduce energy consumption.

Examples are:

1. Establish a compelling energy policy.
2. Implement an energy star procurement policy.
3. Turn off lights and communicate energy conservation plan to all personnel.
4. Unplug appliances and electronics when not in use.
5. Use power chords that turn off when not in use.
6. Verify equipment operational hours and time schedule.
7. Checking illumination levels and switching off excess lighting.
8. Ensure doors and windows are closed as much as possible to prevent heat loss or infiltration.

9. Check door or windows sealant and insulation performance.
10. Conduct an energy awareness campaign that educates the staff, residents, and tenants about their impact on energy use.

1.3 Identifying Energy Waste Brainstorming Sessions:

Identifying energy waste is an excellent strategy. Once you know where and what the energy waste is, it is possible to develop countermeasures that eliminate or minimize them. There are several methods to do so. Re-commissioning and an energy walkthrough audit are the two most known. However, the organization may have to pay a cost to get these done. Some utilities will do free for their customers. In almost every case, these two methods will result in a cost avoidance or savings well above their cost to accomplish. Two cheaper methods can prove excellent at identifying energy waste; they are management/employee brainstorming sessions and energy walkthroughs.

1.3.1 Management/Employee Brainstorming Sessions

First, the team should develop an energy awareness training. Next, the energy team leader, energy manager, or energy champion should accompany a member of top management to the employee/management brainstorming session consisting of a large department's personnel or several small departments or sections.

The senior management representative gives a short speech mentioning the organization's energy goals and the reason and purpose of the brainstorming session that the organization is going to engage its entire people in reducing the energy consumption and cost. Next, the energy awareness training is given by the energy rep that came with the top management rep. After the energy awareness training, the energy rep with the help of a scribe will do the following:

1. Specify clearly the main objective of the meeting, which is related to identifying the potential energy waste in your workplace, for example: "What Energy Waste is Experienced or Evident in your work Area?"
2. Perform "Silent Generation" by having each person identify three energy waste items in their job areas. For example, computer monitors and CPU are not turning off after being idle, the brightness of computers has not been reduced, not using duplex printing, and curtains are over windows not letting light into the room thereby reducing the lumens in the work area.

3. Go "Round Robin" by having the energy rep go around the room and have each participant offer one of their three suggestions and have it written on white board or pad by the scribe. Continue until all possible ideas have been written.
4. Discuss each idea, eliminating duplicates, altering some by consolidation, etc., until a final list is obtained.
5. Normally the ideas are prioritized, and selections are made. However, in this situation, the list is given to the energy team to do the selecting.

These brainstorming sessions are conducted throughout the organization to get inputs and ideas from all the team's personnel. It gets everyone engaged and gives the top management a chance to show their support and commitment. The energy team will consolidate the lists into one main list. The energy team will determine a countermeasure for each idea that will eliminate the waste or at least reduce it. The energy team will track the list until countermeasures have been implemented. Ongoing communications as to progress and results should be provided to the organization's personnel.

1.3.2 Walkthroughs or Energy Audits

Energy walkthroughs are investigations and analysis of facility energy use; it is aimed to identify measures for energy reductions and savings in greenhouse gas emissions. Further, energy walkthrough results in financial benefits by reducing energy consumption. Energy walkthroughs are essential for identifying energy management measures.

To perform an energy walkthrough, several tasks are typically carried out depending on the type of the walkthrough and the size and function of the building. Therefore, an individual energy walkthrough procedure shall be put in place for each facility by itself. Energy walkthrough results in identifying possible energy management measures, and it directs the energy management program to the largest energy use equipment.

The energy team should perform the detailed energy walkthrough and prepare a comprehensive report of findings and recommendations inclusive of feasibility study and return on investment calculations. The report should identify a clear projection of the energy consumption reduction and savings subject to this walkthrough audit.

Saving calculations and energy use reduction should include the following:

- Projections of savings.
- Energy efficiency measures.
- Comparisons with baseline data.

- Tariff rates.
- All anticipated costs for energy efficiency measure with its return on investment.
- A precise time bounded plan for implementation of actions.

Energy walkthroughs are inexpensive and can produce excellent ideas on how to reduce energy use and consumption, provided the team members are experienced in doing energy audits and have facility maintenance and engineering experience.

Purpose: To identify energy waste and determine the appropriate fix.
Who? Facilities, engineers, technicians, energy team leader, and others who can contribute.
What? Kick-off meeting, walk around the building and record anything that uses energy, what it is the amount of energy used (if possible), whether it can reach a state of excessive energy consumption, what preventative maintenance is being performed now, and other pertinent information.

Walkthrough Focused Areas:
Observations:

1. *Occupancy Sensors:*
   Observe infrequently visited areas and determine whether an occupancy sensor will save energy. Look at restrooms, break rooms, copying or printing areas, mechanical areas, hallways, and other areas.
2. *Lights in Administrative Areas:*
   Note types such as T-12s, T-8s, and T-5s. Look for areas daylighting can be used and skylights would help. Look at light bulbs and see if they are dirty with film covering them.
3. *Building Envelope:*
   Search for leaks in doors and windows. Determine if windows should be glazed, caulked, or replaced. Weatherstrip the doors where needed or replace them.
4. *Walls and Roof Insulation:*
   Check the insulation level and determine if more would help.
5. *Motors and Other Equipment Except for HVAC:*
   Note each and check the switches and sensors associated with each. Check time schedule of each equipment and whether it runs according to it or continuously running.
6. *Data Centers:*
   Look for hot and cold aisles and whether hot air is kept from commingling with the cold air on its return to the computer (CRAC).

7. *Security Lights:*
   Check to see if they are adequate and energy friendly.
8. *HVAC:*
   Note brand, capacity, date installed, the motors, and switches associated with the system, and check roof vents and other parts for adequacy and maintenance.
9. *Building Automation System (BAS) and Metering:*
   See if BAS is outdated. Note where additional metering can help identify potential problem areas.
10. Computers, monitors, imaging equipment, fax machines, and other office equipment.

Walkthroughs can also be done along with ECM when the machine is selected to be in ECM or when determining the significant energy users to comply with ISO 50001 Energy Management Systems (EnMS). The walkthrough results will be placed into low hanging fruit, energy conservation, energy efficiency, and energy centered maintenance programs for resolution and energy consumption reduction.

1.4 Energy Conservation

Energy conservation refers to reducing energy consumption by using less of energy input. Energy conservation is different from efficient energy use, which is using less energy for the same or more output. Driving for less time is an example of energy conservation while driving a vehicle that gets more mileage per gallon is an example of energy efficiency. Turning out the lights when not in use, unplugging appliances or electronics when not in use, making your computer monitors go to sleep after a period of idle time, and using duplex printing when possible are examples of energy conservation items. All organizations wishing to reduce energy conservation should develop and implement an energy conservation program and recognize and reward success. A simple, but effective, energy conservation program is shown in figure 1.2 below.

1.4.1 Step 1: Create an Energy Conservation Program

Put the energy conservation program into a power point presentation. Include the typical items such as:

1. Switch off the lights when space is not used.
2. Unplug appliances, equipment, and electronics when you leave the room.

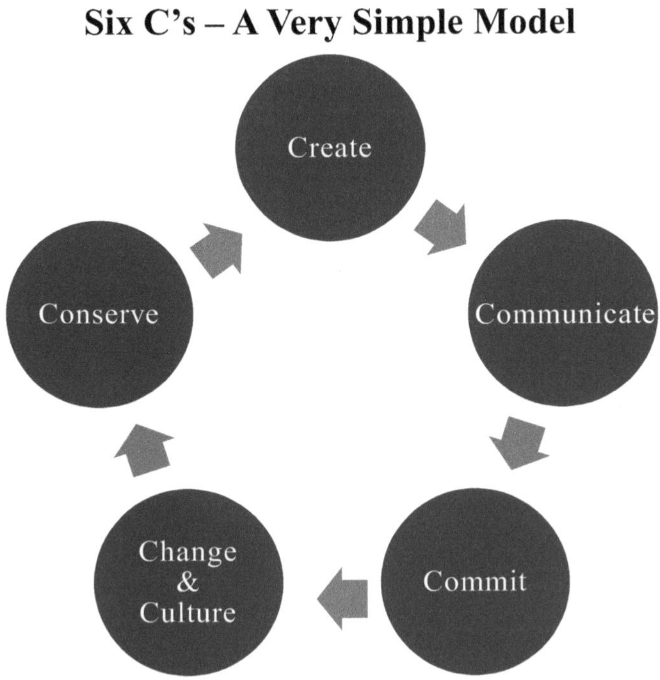

Figure 1.2 A simple, but effective, energy conservation program.

3. Purchase energy star equipment, appliances, and electronics. It is recommended that all organizations make this a procurement policy.
4. Set computers to turn off after 15–30 minutes of non-use (hibernate or system standby).
5. Set monitors to go to the sleep function after being idle for 15–30 minutes.
6. Turn down the brightness of the computer monitors and televisions.
7. For small refrigerators: Put a bag of ice in the refrigerator, clean the coils periodically, and unplug when on long holidays and vacations.
8. Use power chords that turn off when not in use.
9. When printing or copying, use 30% or higher recycled paper.
10. Use duplex printing when feasible.
11. If practical, use network printing instead of everyone having a printer.
12. Use electronic files when possible. Do not keep paper backup unless required by headquarters.
13. Put on your emails: "Do Not Copy Unless Necessary. Save Paper."
14. Report any energy or water problems to facility management.

15. Use a CFL bulb or LED fixture for task lights.
16. Actively participate in alternative work programs.
17. Wear proper clothing when ASHRAE temperature settings are set for cooling and heating. Do not tamper with the thermostats.
18. At work or on campus, take short showers (2 minutes recommended).
19. Use only the paper towels when needed. Organization should use only heavier paper towels so that four or five papers do not come out at a time.
20. Check faucets in your area for leaks. Report any leaks to facilities management.
21. Use the stairs for first two floors going up and first three floors coming down.
22. Only use a cold wash.
23. Set hot water heaters to 120°.
24. Only use dishwashers for full loads.
25. Use blinds or curtains in all areas to help maintain comfort.
26. Keep windows and doors closed when air conditioning or heat is provided.

1.4.2 Step 2: Communicate

Develop an energy conservation presentation to be given by your energy champion or manager to all personnel. The presentation should include the company or organization's energy reduction goal and the definition of energy efficiency and energy conservation. Providing energy conservation training including getting everyone to commit and do the 26 items above is our simple, but effective, energy conservation program. Experience has shown that implementing the 26 energy conservation items can reduce our energy consumption by 5%–8%. All personnel should receive this energy conservation training and receive a certificate showing they did, and it should go into their individual training records.

1.4.3 Step 3: Commit

During the energy conservation training, have all participants commit to practicing energy conservation. A good way to do this is on earth day, April 22 each year; have either individual commitment sheets or put the 26 items on a large sheet and have the ones that voluntarily commit by signing the large sheet. They are committing for the next year that they will support and do energy conservation items to help the organization achieve its energy reduction goals. Why should they do this? Because it is just the right thing to do. It helps the individual, the organization, and our planet.

1.4.4 Step 4: Changes and Culture

To have a successful energy conservation program, people often have to change their behaviors to participate. Most people want to contribute but may not know what to do. The training will correct this problem. Some who knows what to do, do not do it. Behavior training will help enforce this type of behavior. Chapter 14, Building Energy Centered Behavior leading to an Energy Centered Culture, will cover more about achieving the desired behavior.

1.4.5 Step 5: Conserve

Steps 1–4 help an organization to implement energy conservation which results in conserving electricity. To sustain the program, the positive results must regularly be communicated through speeches, staff meetings, or on a dashboard. Most importantly, individuals who contributed and made significant contributions are recognized and rewarded.

1.5 Energy Efficiency Projects

The Department of Energy and Climate Change defines energy efficiency as "a measure of energy used for delivering a given service. Improving energy efficiency means getting more from the energy that we use." **Doing the same but with less energy or energy efficiency is doing the same mission with less energy**.

Examples of energy efficiency projects are:

1. Upgrading HVAC.
2. Putting double paned windows in the facility.
3. Replacing the UPS for a data center.
4. Replacing and modernizing the cooling tower.
5. Putting panels in a data center to block the cold air returning to the inside computer (commuter room air conditioner (CRAC)) and mixing with the hot air.

Energy efficiency projects can significantly reduce energy consumption. The walkthroughs can help identify energy efficiency projects. Next, the payback (the time the savings will pay for the project) is computed or calculated and projects with a payback of 3 years or less will most probably be funded by the organization. The government will consider plans up to 11 years' payback to reduce energy consumption. Energy centered maintenance is an energy efficiency measure. An organization should have a mixture of energy conservation, energy efficiency, and low hanging fruit items implemented in their energy reduction program.

1.5.1 Energy Objectives and Targets with Action Plans

A best practice is to have a cross-functional energy team representing all the major organization's departments. They will need to be trained in energy awareness, process improvement, problem solving and development of objectives, targets, and action plans.

There are three types of goals. The first is one to improve something such as reducing energy consumption, increasing the number of suggestions or ideas, or development of a new procedure. The second purpose is to maintain something you desire not to degrade such as environmental performance. The third type of objective is to determine if something is feasible or not. Feasibility of going on time in use energy, accepting demand response, and changing water heater from gas to solar are examples of this type of objective. During the first year, several of the objectives will be developing something such as energy policy, energy plan, energy procedures, development of energy awareness training, and others. Then emphasis on reducing energy consumption objectives and targets will be prevalent. Installing occupancy sensors in restrooms, break rooms, mechanical rooms, and other areas with occasional visitors throughout day or night shift and reducing office paper use and implementing IT power management are good examples.

Objectives of Energy Reduction:
The overall purpose of energy reduction is to achieve and maintain optimum energy purchase and utilization throughout different consumer types, such as factories, commercial and residential developments, etc.

The implementation of an energy management process reflects the organizationally responsible behavior in preserving natural resources, reducing the impact on the environment, reducing greenhouse gas emissions, improving air quality, and limiting global climate change.

Therefore, the implementation of an energy reduction program will result in the following objectives:

- Enhance energy efficiency continuously by implementing an effective energy management program that supports all operations and achieves customer satisfaction while providing a safe and comfortable environment.

- Developing and maintaining effective monitoring, reporting, and management strategies for wise energy consumption.
- Finding new and better ways to increase returns on investments through research and development and energy saving initiatives.
- Developing interest and dedication to the energy management program from all building's operators, employees, tenants, shareholders, owners, and visitors.
- Reducing operating expenses and increasing asset values by actively and responsibly managing energy consumption.
- Reducing greenhouse gas emissions, mainly CO_2 emission and reducing carbon footprint, caused by energy consumption.
- Complying with regulatory laws and legislations listed by the government.
- Support the growth of renewable energy resources and sustainability commitments.

1.5.2 Characteristics of a Successful Energy Reduction Program

Eight characteristics keep showing up at organizations that have been successful at reducing energy consumption and energy costs. They are:

1. Top management leadership supports, committed and involved in the energy reduction effort and becomes the program GLUE (**G**ood **L**eaders **U**sing **E**nergy).
2. Energy |reduction is made a corporate priority.
3. Corporate goals are established and communicated.
4. The energy champion or energy manager or both along with their cross-functional energy team select challenging strategies that include development of an energy plan and objectives and targets with actions plans.
5. Both key performance indicators (KPIs) and key result indicators (KRIs) are employed and kept current and visible to measure and drive progress and results.
6. Sufficient resources are provided to fund or ensure that adequate countermeasures are implemented to achieve the corporate goals.
7. An energy centered culture is achieved.
8. Sufficient reviews are conducted to ensure that continuous improvement, compliance to legal requirements, and adequate communications

are provided to keep all stakeholders informed, motivated, and engaged.

For these eight characteristics or success factors, there are best practices that enable them to stand out as crucial to obtaining success. For the first success factor, we have:

- **Top management leadership** supports, committed and involved in the energy reduction effort and becomes the programs GLUE (**G**ood **L**eaders **U**sing **E**nergy). Appoints an energy champion or energy manager who establishes an X-F energy team.
    - Shows commitment in the energy policy and communicates the policy, goals, success stories, and how everyone can support the program.
    - Provides support and resources and monitors energy reduction.
    - Programs and leads annual executive review.
- **Energy reduction** is made a corporate priority.

Top management communicates this in the energy policy. Senior managers give motivating and informative speeches on reducing energy consumption and why it is important. Actions show management commitment.

- **Corporate goals** are established and communicated.

Energy reduction goal is established. Renewable energy goal(s) are set. They could be multi-goals – first two years short-term, middle-term 10 years, and neutrality achieved long-term goal. Other sustainability goals are set. All goals are communicated to all involved. Greenhouse gas emissions reduction goal(s) is established.

- **The energy champion** or energy manager or both along with their cross-functional energy team select challenging strategies that include development of an energy plan and objectives and targets with actions plans.
    - Perform walkthroughs or have an energy audit and re-commissioning.
    - Conduct management/employees brainstorming sessions.
    - Appoint functional teams.
    - Perform energy research.
    - Encourage employee suggestions.
    - Develop and train everyone in energy awareness, energy conservation, and energy efficiency.
    - Conduct monthly energy team meetings that are effective.
    - Use critical success factors (CSFs) to measure progress and drive increased performance.

  - Use the energy reduction checklist to identify areas to improve.
  - Use both key performance indicators (KPIs) and key result indicators (KRIs). Use the KPIs to drive the KRIs that measure the goals.
  - Use data collection forms for each indicator.
  - Design and use dashboards if possible.
- **Sufficient resources** are provided to fund or ensure that adequate countermeasures are implemented to achieve the corporate goals.
  - Develop and implement an energy conservation program.
  - Get everyone to commit.
  - Identify energy efficiency projects with excellent payback.
  - Implement energy center maintenance (ECM) in buildings, manufacturing facilities, universities, and industrial buildings, including data centers.
  - Implement the low hanging fruit items.
  - Estimate each as to its contribution and have contributions equal to or greater than the goals.
- **An energy centered culture** is achieved. Organization's values and principles are identified and taught.
  - Energy training provided.
  - Committed and caring leadership is apparent.
  - Desired behaviors are encouraged and training is provided to ensure what is desired is known by participants.
- **Sufficient reviews** are conducted to ensure that continuous improvement, compliance to legal requirements, and adequate communications are provided to keep all stakeholders informed, motivated, and engaged.
  - Executive or management reviews are held at least annually with the required inputs and outputs.
  - The energy team conducts a legal compliance review to assess whether all legal requirements are being achieved.
  - Minutes are kept and made available for interested parties to review.

The above is an overview of implementing a successful energy reduction program at any organization, business, university, or college. Energy centered maintenance (ECM) is an effective energy efficiency strategy. Its development including history, steps to implement, benefits, examples such as reducing energy consumption in data centers and manufacturing, measuring the efficiency and effectiveness, finding root causes and fixing them, and other pertinent information will be presented in the following chapters.

1.5.3 Energy Centered Maintenance

Energy centered maintenance (ECM) is a continuous improvement maintenance regime that combines the physical preventive and predictive maintenance tasks with energy-related maintenance tasks that maintain the operational parameters of the equipment and its efficiency (i.e., motor current or fan flow rate).

The primary purpose is to reduce energy use by identifying equipment or items that can become energy hogs while still performing their function and prevent that from occurring.

Energy centered maintenance supports the energy reduction program adopted within the facility and helps achieving the projected energy savings.

Why is ECM needed?

- Poor maintenance of energy-using systems, including significant energy users, is one of the leading causes of energy waste in the Federal Government and the private sector.
- Energy losses from motors not turning off when they should, steam, water and air leaks, inoperable controls, and other losses from inadequate maintenance are large.
- Uses energy consumption excess or energy waste as the primary criterion for determining specific maintenance or repair needs.
- Lack of maintenance tasks in measuring the operational efficiency of the equipment such as motor power consumption and equipment effectiveness.

The walkthroughs or energy audits, the management/employees brainstorming sessions, energy conservation items and energy efficiency projects, implementing low hanging fruit items, and implementing energy centered maintenance are excellent ways to reduce energy waste in your organization, business, or university. There are a few administrative moves that will help the energy waste reduction strategy. First, have an energy champion or energy manager to run the day-to-day energy-related activities and to coordinate with the maintenance team to ensure that the maintenance activities are conducted effectively. He or she can benefit by having a cross-functional energy team with representatives from all the main departments including one or two members from facilities and one from engineering. For large organizations, there may be several other teams that either report or inform the energy manager and seek his support and guidance. On a university campus, you can have energy conservation teams for large buildings. And to assign operational teams from major departments to identify waste and develop countermeasures that minimize or eliminate the energy waste that is resulted from lack of proper maintenance.

The ECM planning, designing, and implementing efforts can be accomplished by one or two engineers or by a maintenance leader and a small team of 4–5 members.

Steps in Implementing ECM:

The following steps should help the energy management and maintenance teams in implementing the ECM policy within the organization.

The details of the ECM model steps are discussed in depth in the subsequent chapters of this book.

Some steps will enable ECM implementation to be accomplished efficiently and effectively.

Step 1. Obtain top management approval and commitment to energy centered maintenance (ECM):

- Make them aware of ECM purpose, objectives, and concept.
- Estimate ECM's contributions to the organization's energy reduction plan and reducing greenhouse gas emissions.
- Start developing key contacts in other departments that can help ECM become a reality such as operations, logistics, environmental, facility maintenance, and others.

Step 2. Identify the equipment and systems most likely to use excessive energy:

- Make a list of these systems and equipment and then prioritize them.

Step 3. Determine what systems will be needed to track the ECM activities:

- Consider all systems that are major contributors to energy consumption and determine which equipment are most energy consuming.

Step 4. Pilot a potential energy hog system:

- Commit to addressing at least one of these troubled energy hog systems for validation of ECM or as a pilot where value can be shown and proven and baseline information can be developed.
- Begin baselining/tracking this system:
  - System operations and history.
  - System maintenance and history.
  - System costs, time to service, downtime, resulting over time, OEE, machine efficiency, etc.

Step 5. Determine what ECM tasks need to be conducted:

- Consider required tasks, skill set requirements, tools and equipment, cost effectiveness, etc.

Step 6. Determine what proactive measures should be included in the regular maintenance plan:

- Consider purchasing or enhancing a computerized maintenance management system and commit to its implementation and use and update it to include the ECM tasks.

Step 7. Purchase diagnostic, volt meters, amp meters, other metering, or monitoring equipment necessary for ECM:

- Be sure not to purchase an equipment or items for inspections if the organization already has them.

Step 8. Achieving maintenance and operational efficiency:

- Understand how to operate this system correctly:
  - Define and complete operator training needs.
- Understand how to maintain this system correctly:
- Define and complete maintenance training needs and establishing specific inspection procedures, what to look for, and maintenance tasks.

Step 7. Train appropriate personnel in ECM purpose, concept, benefits, and how it fits into present maintenance program:

- Train Maintenance Manager, Supervisors, and Technician Leads

Who: Maintenance Manager, Maintenance Supervisors, Technician Leads or Foremen, Technicians, or Mechanics.

What:

1. ECM purpose, objectives, and concept.
2. Seven types of maintenance – advantages and disadvantages.
3. Inspections – What to look for, tools needed, and data needed.
4. Equipment identification codes.
5. Updating PM plans.
6. Sample problem, cause, effect, and corrective preventative problem.

Step 8. Train Technicians or Mechanics in ECM procedures:

- Equipment identification.
- Inspections requirements.
- CMMS and data requirements.
- Maintenance, repair, or replacement decisions.

ECM Operating Principles:

Five operating principles guide ECM. They are:

1. Find waste and eliminate it.

2. Perform quality inspections and maintenance or replacement.
3. Be both efficient and effective.
4. The Maintenance Program Management Team continually analyzes the maintenance data to identify trends, inefficiencies, and to develop strategies for operational and financial improvements. Continuous improvement is a goal.
5. ECM addresses and solves all related environmental requirements, concerns, or issues.

Some of the five are self-explanatory but not 3, 4, and 5.

1.5.4 Be Both Efficient and Effective

Someone once stated, it is possible to be efficient but not effective, or productive but not efficient. Our goal should be that the maintenance technicians are both efficient and effective. For example, operations release Machine XYZ to support to test for excessive energy use on Wednesday morning. They want the machine back on Thursday morning by 10:30 am. The maintenance technician checks the Volts and Amps and found that both exceeded their nameplate amounts. The machine's bearings were grinding and needed to be replaced. Inventory Management only had one-half of the bearings needed and had to order from their central parts store area the remainder. Although they had them on special order, they came in overnight arriving at 8:20 pm. The technician picked them up from the parts delivery person and rushed to the machine. He quickly installed them and had the equipment running smoothly using only nameplate power by 9:50 am. The Operations Required Date was met, so the job had been useful but was not sufficient since the technician had to travel to the machine site more than once and did not have the parts needed. The previous sentence is an example of "Being Effective But Not Efficient." What about being efficient but not effective? Let us say another technician was assigned to do an ECM inspection of machine HWX that was released to maintenance by operations for only 2 hrs, starting at 8:45 am. Operations team were put under pressure due to the short notice interval to do the inspection, and conduct some maintenance. The technician did the inspection and found that excessive energy was being used. He found the trouble or root cause right away and started the job tasks immediately. The work that had to be accomplished was huge and required more than one technician. They completed the work at 11:00 am within the man-hour estimate or standard but were 15 minutes late to operations, thus delaying production 15 minutes of unplanned downtime. The technicians were efficient but not effective.

The maintenance technicians have to consider operations as their customer. The Operations Required Date is their customer's requirement, and they need to meet it. However, in the past, one of the authors has seen them act like they are each other's enemy since maintenance felt the operators set non-realistic requirements and the operators did not think maintenance planned the job well and took unnecessary breaks or were unproductive sometimes. They must work together as a team respecting each other and honoring each need.

1.5.5 Analysis of the Maintenance Data and Continually Improve

The computerized maintenance management system (CMMS) provides a lot of meaningful data: the number of job plans completed on time; the percent of the time that targeted machines have excessive energy use; the number of times the equipment could be fixed with just maintenance or had to be repaired or replaced; the times needed to do the inspection and the maintenance so that historical standards could be developed or compared to the present standards to be sure they are realistic. Data should be trended, and the trends analyze to learn what is happening, what is going well, and what should be changed.

Continuous improvement should occur if the Plan-Do-Check-Act Wheel is followed. Job plans are developed, applied by the technicians doing the inspections and maintenance, recorded by CMMS, and then provided to the maintenance analyzed to check if the job plan was okay or needed change.

ECM Addresses and Solves All Related Environmental Requirements, Concerns or Issues:

Today, both operations and maintenance must be environmentally aware and manage the environmental aspects that can impact the environment. Maintenance typically deals with numerous environmental aspects that could impact the environment and the workplace. To name a few: grease, oil, lubricants, cleaning supplies, aerosol cans, batteries, electronic equipment, refrigerants, flammable materials, fuel, bulb disposal, acids, paint, ether glycol, foam, packaging, and compressed gas. Each of these aspects can impact the workplace or the environment. Some of them have legal requirements that must be dealt with to ensure legal compliance. The aspects vary on the risk as to the seriousness of impact, the probability of impact, and whether an operational control has been designed to help minimize the impact. They also use and must deal with hazardous materials which must be managed effectively to ensure compliance with federal and state laws. If not, fines can be levied, and the government can close your workplace due to non-compliance environmental problems.

All maintenance technicians or mechanics and their supervisors must be trained on environmental aspects that could impact the workplace – issues such as lubricants, acids, oil, and other items getting into the ground water, what to do in case of an environmental spill, what to do with aerosol cans, how to store flammable materials, and other environmental areas. Each supervisor, technician, and mechanic must become an environmental steward and help keep their workplace safe and productive.

ISO 14001 Environmental Management System is one of the best ISO standards written. It provides excellent guidelines and procedures to enable any organization to properly management its aspects in their workplace. By adopting the ECM model, the energy and maintenance teams ensure that they are minimizing the impact of equipment operation on the environment. Minimizing the energy waste during equipment operation will reduce the generation of greenhouse gases.

2

Different Maintenance Types and the Need for Energy Centered Maintenance

2.1 History of Maintenance

It is hard to say exactly when maintenance started. During the industrial revolution, some equipment maintenance was accomplished. Facilities maintenance for years was normally emergency maintenance such as a break in the water line, electricity outage in the building, a leak in the roof, or a broken window. Those examples came to be called breakdown or reactive maintenance. Preventative maintenance came in 1951, followed closely by periodic maintenance, and predictive maintenance, all three main maintenance programs. They were defined, procedures were written, and maintenance planning and maintenance records became a reality. Non-emergency tasks such as painting, lubricating, replacing bearings, replacing burned-out light bulbs, repairing door locks, caulking around windows, replacing filters, etc., became part of the maintenance of all the main buildings since management now believed that was the most cost-effective way of preserving their investment. Predictive maintenance was valued in the manufacturing facilities in that machine efficiency and personnel productivity benefitted from this new predictive capability.

In 1978, total productive maintenance (TPM), invented by the Japanese as part of their total quality improvement effort, was introduced by Toyota in a widely acclaimed book on Toyota's achievements in quality improvement. TPM got the employees engaged and followed the simple concept "Everything has a Place and Everything in its place." Cleaning the workplace was one of the first requirements of TPM. The concept also uses the six S's technique to get everything in its place, standardized and sustained. About this time, a new concept called reliability centered maintenance (RCM) came along and simply stated that all equipment are not equal.

Sometimes we need to perform regular proactive maintenance on some equipment, and have some other equipment run to fail based on our operational strategy. RCM included using statistics and determining root causes, thereby limited its application.

In 2012, a common strategic goal or objective in most large businesses, universities, and organizations was to reduce energy consumption which saved money and contribute to reducing greenhouse gas emissions. Energy as an aspect in ISO 14001 environmental management system (EMS) was not receiving sufficient attention; so ISO 50001 energy management system (EnMS) was created. Maintenance managers felt energy and maintenance were related but had not defined how. Motors running when they should not, machines running while using excessive electricity, servers using excessive energy, and other similar happenings resulted in energy centered maintenance being born in 2012. Although seminars on ECM have been given since 2013, no book has been written. This book is the first book written on ECM to enable energy savings from maintenance practices to be realized.

2.2 The Maintenance Types

There are seven recognized maintenance types counting energy centered maintenance. They are:

1. Breakdown or reactive maintenance (Before 1950, manufacturing revolution).
2. Preventative maintenance (1951).
3. Periodic maintenance (1951).
4. Predictive maintenance (around 1951).
5. Total productive maintenance (1951 origin; 1980s in the USA).
6. Reliability centered maintenance (1960s origin; 1978 became known).
7. Energy centered maintenance (2012).

The reference "htpps://www1.eere.energy.gov/femp/pdfsOM_5.pdf" available on the Internet will be used since it best describes the first six maintenance types.

2.2.1 Breakdown or Reactive Maintenance

Basic philosophy:

- Machinery runs until it fails.
- Repair or replace damaged equipment only when problems occur including failure.

Reactive maintenance is often thought of as "run it till it breaks" maintenance strategy.

No maintenance tasks are taken to maintain the equipment as the manufacturer originally intended.

Breakdown or reactive maintenance can be defined as activities are also known as "run-to-failure." Using this approach, maintenance will be performed just when the asset's deterioration causes a functional failure. Reactive maintenance is ideally used when the failure does not significantly affect operation, production, or generate any financial losses other than repair costs if the financial impact is less than the cost of preventing the failure.

Advantages:

- Low cost.
- Less maintenance personnel.

Disadvantages:

- Additional cost due to unplanned downtime of equipment.
- Could increase labor cost if overtime is required.
- Cost of repair or replacement of equipment.
- Could have secondary equipment or process damage from equipment failure.
- Inefficient use of maintenance personnel.

2.2.2 Preventive Maintenance (so called, time-based maintenance)

Basic philosophy:

- Schedule maintenance activities at predetermined time intervals.
- Repair or replace damaged equipment before obvious problems occur.

"Preventive maintenance can be defined as actions performed on time- or machine-run-based schedule that detects through inspection and prevents degradation of a component or system with the aim of sustaining or extending its useful life through controlling wear and tear to an acceptable level."

The proven benefit of preventive maintenance is that it provides the first level of control of maintenance beyond the reactive level to prevent failures. Carefully managed preventive maintenance minimizes corrective maintenance, yet it carries risk. The maintenance intervals chosen for the "life" of the asset may not have a strong statistical base, making the likelihood of over-maintaining or under-maintaining the asset very high.

Maintenance activities generated from preventive maintenance and/or predictive maintenance are referred to as corrective maintenance. These activities should be planned and scheduled in advance of a failure.

htpps://www1.eere.energy.gov/femp/pdfsOM_5.pdf

Advantages:

- Cost-effective in most organizations (12%–18% cost savings over reactive maintenance).
- Very flexible.
- Achieve some energy savings.
- Reduced equipment failure.

Disadvantages:

- Catastrophic failures will likely occur.
- Labor cost is high.
- Includes performance of some maintenance that may not need to be performed.

htpps://www1.eere.energy.gov/femp/pdfsOM_5.pdf

2.2.3 Predictive Maintenance (Condition-Based or Time-Based Maintenance)

Basic philosophy:

- Schedule maintenance activities when mechanical or operational conditions warrant.
- Repair or replace damaged equipment before obvious problems occur.

"Predictive maintenance can be defined as follows: Measurements that detect the onset of system degradation (lower functional state), thereby allowing causal stressors to be eliminated or controlled before any significant deterioration in the component physical state. Results indicate current and future functional.

The technical basis for predictive maintenance is that most ailing assets provide sufficient warning of the fact that they are in the process of failing during operation. Predictive maintenance replaces preventive maintenance with maintenance scheduled only when the asset requires in advance of functional failure, enabling maintenance resources to be efficiently planned.

capability." htpps://www1.eere.energy.gov/femp/pdfsOM_5.pdf

Advantages:

- Increased component life cycle.
- Allows for corrective actions. Otherwise, the component would fail.

- Decrease in equipment downtime.
- Decrease in parts and labor costs.
- Improved worker and environmental safety.
- Improved workplace climate and worker morale.
- Some energy savings.
- Estimated 8%–12% cost savings over preventive maintenance program.
- Reduction in maintenance costs: 25%–30%.
- Elimination of breakdowns: 70%–75%.
- Reduction in downtime: 35%–45%.
- Increase in production: 20%–25%.

https://www1.eere.energy.gov/femp/pdfs/OM_5.pdf

Disadvantages:

- Increased investment in diagnostic and vibration analysis.
- Increased investment in maintenance personnel. "Predictive maintenance differs from preventive maintenance by determining maintenance needs on the actual condition of the machine or equipment rather than on some preset schedule."
- Remember that preventive maintenance is time-based. Actions such as changing oil are based on time, like calendar time or equipment run time. For example, most people change their oil in their cars every 3000–5000 miles.

https://www1.eere.energy.gov/femp/pdfs/OM_5.pdf

2.2.4 Reliability Centered Maintenance

Basic philosophy:

A structured/logic-based process is used to develop complete system and equipment maintenance programs providing the highest level of equipment reliability at the best possible cost.

It utilizes predictive/preventive maintenance techniques with root-cause failure analysis to detect and pinpoint the precise problems, combined with advanced installation and repair techniques, including potential equipment redesign or modification to avoid or eliminate problems from occurring.

The reliability centered maintenance (RCM) magazine provides the following definition of RCM: "a process used to determine the maintenance requirements of any physical asset in its operating context."

RCM recognizes that all equipment in a facility is not of same importance to either the process or facility.

It recognizes that equipment design and operation differ and that different equipment will have a higher probability of undergoing failures than others. RCM will enable a facility to more closely match their resources to maintenance needs while improving reliability and decreasing cost.

Advantages:

- Lower costs by eliminating unnecessary maintenance or overhauls.
- Reduced probability of sudden equipment failures.
- Enables maintenance personnel to focus maintenance activities on critical components.
- Increased component reliability.
- Incorporates root-cause analysis.
- Improves reliability and availability.
- Reduce downtime.
- Reduces maintenance costs.
- Improves quality
- Improves Uptime
- Improves profitability

Disadvantages:

- Can have large start-up costs, extensive training, etc.
- More difficult to sell to management since it has more root-cause and other analysis as part of its implementation methodology.

RCM Maintenance Priority:

1. Emergency life, health, safety risk-mission criticality.
2. Urgent continuous operation of facility at risk.
3. Priority mission support/project deadlines.
4. Routine prioritized: first come/first served.
5. Discretionary desired but not essential.
6. Deferred accomplished only when resources allow.
   htpps://www1.eere.energy.gov/femp/pdfsOM_5.pdf

2.2.5 Total Productive Maintenance (TPM)

Basic Philosophy:

It includes equipment maintenance including autonomous, planned, and quality maintenance.

It includes training and education and the application of the 5 S's – Sort-Straighten-Shine-Standardized-Sustain.

"In industry, **total productive maintenance** (TPM) is a system of maintaining and improving the integrity of production and quality systems through the machines, equipment, processes, and employees that add business value to an organization."

https://www.google.com/?gws_rd=ssl#q=Total+Productive+maintenance

"TPM (total productive maintenance) is a holistic approach to equipment maintenance that strives to achieve perfect production:

- No breakdowns.
- No small stops or slow running.
- No defects.
- Also, it values a safe working environment.
- No accidents.

TPM emphasizes proactive and preventative maintenance to maximize the operational efficiency of equipment. It blurs the distinction between the roles of production and maintenance by placing a high emphasis on empowering operators to help maintain their equipment.

The implementation of a TPM program creates a shared responsibility for equipment that encourages greater involvement by plant floor workers. In the right environment, this can be very effective in improving productivity (increasing up-time, reducing cycle times, and eliminating defects)."

Advantages:

- Gets production and maintenance working together.
- Gets operators involved.
- Uses 5 S's.
- Has focus.

Disadvantages:

- Training costs.

2.2.6 Energy Centered Maintenance

Basic Philosophy:

1. Find It-Fix It.
2. Repair or replace when equipment is pulling excessive Amps or Volts.
3. Replace equipment that is an energy hog.
4. Operate equipment in a shift only for the time needed.
5. Maintain the equipment to ensure operational performance of the machine is functioning efficiently as intended.

Energy centered maintenance is defined as a continuous improvement maintenance regime that combines the physical preventive and predictive maintenance tasks with energy-related maintenance tasks that maintain the operational parameters of the equipment and its efficiency (i.e., motor current or fan flow rate).

The energy centered maintenance model applies to all energy critical equipment to extend its life, maintain equipment efficiency, prevent excess energy use, and reduce energy waste.

Advantages:

- Saves significant energy.
- Keeps equipment from failure.
- Prolongs equipment life cycle
- Ensures equipment is operating according to the intended design function.
- Increases energy efficiency of the equipment through a low operation and maintenance cost.
- Can be integrated easily with planned preventive maintenance.
- Identifies improvements which can be made to increase equipment's efficiency.

Disadvantages:

- Have to find the equipment with the condition.

Maintenance Strategy:

An effective maintenance approach results from a well-organized and carefully executed effort by the maintenance team. They identify and define the maintenance tasking and use established standard failure codes of the balanced proactive approach. The maintenance staff takes input from several sources, including facility/account management, facility operations, and the maintenance staff.

A critical system does not imply that all maintenance tasks on it prevent failure. Each facility, system, or piece of equipment should be examined/classified by its impact on facility operation to determine the criticality of the equipment operation on the facility business. To determine the maintenance approach, measure the facility, systems, and equipment against the following criteria:

- High repair/replacement costs.
- High energy consumption/conservation.
- Effect on the facility value.

The method for identifying the maintenance approach by maintenance management is summarized in the following steps:

- ○ Consider the facility goals and requirements for system operation.
- ○ Examine the facility and identify assets whose proper performance is necessary to meet organization business and maintenance requirements.
- ○ Determine the impact (of the functional failure of each facility asset and equipment.
- ○ Determine what predictive and/or preventive tasking will mitigate potential failures.

If the cost of such maintenance actions is less than the impact of the functional failure, assign the appropriate predictive and/or preventive maintenance activity on the maintenance schedule.

The maintenance strategy involves tasks that can be applied to prolong the useful life of equipment and prevent/avoid premature failures.

The strategy for selecting the appropriate maintenance approach involves the following:

- Consider the criticality of the equipment on facility operation and determine what kind of maintenance approach is appropriate, such as run to failure, preventive maintenance, predictive maintenance, total productive maintenance, and reliability centered maintenance.
- Consider the variety of problems (failures) that may develop in equipment.
- If preventive maintenance approach is not adequate to detect the variety of maintenance problems, use predictive maintenance or reliability centered maintenance approaches.
- If predictive maintenance or reliability centered maintenance does not adequately apply or is cost prohibitive, use preventive maintenance. Develop inspection tasks to reveal the failures not adequately covered by predictive maintenance and reliability centered maintenance.
- Decide the combination of approaches (predictive/proactive) to develop a total productive maintenance approach and then determine the frequency of the particular task. Usually, a combination of approaches provides the required coverage to assure reliable performance.
- Development of maintenance strategy will help the maintenance team to:
  - ○ Concentrating maintenance resources where they will do the best.
  - ○ Performing technically efficient and cost-effective maintenance.
  - ○ Eliminating unnecessary and ineffective maintenance tasks.
  - ○ Developing a documented basis for the maintenance program.

2.2.7 Detection of Potential Failures

The detection of potential failures in the maintenance strategy yields an important means of improving the overall maintenance program. Early detection of impending problems prevents a critical level of deterioration or failure; so the maintenance organization can use lead time to deliberately plan repair or improvement work and then carry it out using resources more efficiently.

The maintenance strategy is based on the fact that many failures do not occur instantaneously but develop over a period of time. If evidence that a failure is occurring can be found, it may be possible to take action to prevent the failure and/or avoid the consequences.

The detection of potential failures approach focuses on the failure interval, allowing the maintenance organization the opportunity to use the time to deliberately plan repair or improvement work and then carry it out using resources more effectively.

The principal aim of the maintenance approach is to uncover problems or manageable effects before they reach the crisis stage of equipment failure or breakdown. The sooner the problems or effects are found, the greater the opportunity for planning, gathering materials, coordinating outages, estimating, and allocating resources.

3

Energy Centered Maintenance Origin and Model

3.1 Origin of ECM

Years ago, after giving a speech on maintenance at an Industrial Engineering Conference, I was asked: "Is there any connection with maintenance and saving energy." I had not been asked this question before. My answer is yes. If we paint a room white, it will take fewer lumens to serve the place. If we change a filter in an air conditioner on time, the air conditioner will not have to operate with higher airflow resistance, thus saving electricity. That question in the last 15 years has often been asked. Organizations that provide facility and equipment maintenance would like to advertise or be able to tell their client that they save energy while providing their maintenance. They were hesitant until ECM came along. ECM shows a direct relationship between reducing energy consumption and performing maintenance tasks. They knew that:

- Poor maintenance of energy-using systems, including significant energy users, is one of the leading causes of energy waste in the Federal Government and the private sector.
- Energy losses from motors not turning off when they should, steam, water and air leaks, inoperable controls, and other losses from inadequate maintenance are large.
- Uses energy consumption excess or energy waste as the primary criterion for determining specific maintenance or repair needs.
- Lack of maintenance tasks in measuring the operational efficiency of the equipment such as motor power consumption and equipment effectiveness.

ECM uses energy consumption excess or energy waste as the primary criterion for determining specific maintenance or repair needs. Energy centered maintenance (ECM) was originated in 2012 when Marvin Howell

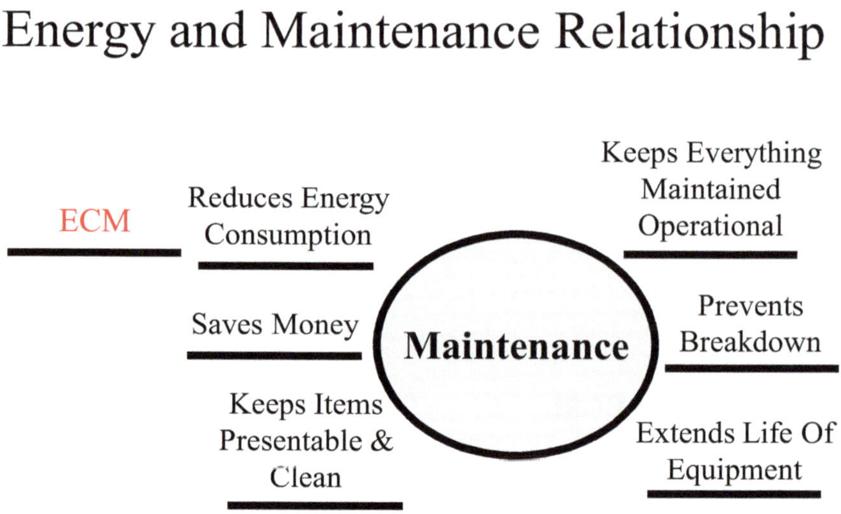

Figure 3.1 Maintenance and energy relationship.

kept finding motors running 24/7 when they were only required to run 7–8 hrs daily. Also, he observed switches stuck on equipment, sensors not working, building automation systems with operators not trained, data centers using servers that were energy hogs, and cold air mixing wrongly with the hot air on the way to the computer room air conditioner (CRAC). He discovered that a maintenance program needed to address equipment using excessive energy. Of the seven different maintenance systems, none addressed this energy waste as the primary focus. Figure 3.1 shows the relationship. Marv was excited to present this new concept on Association of Energy Engineers (AEE) On-Line Energy Seminars.

3.2 The Model – Its Aim and Design

In 2015, Fadi Alshakhshir developed and designed a detailed process of implementation of the energy centered maintenance (ECM) model. The aim of the model is to introduce new maintenance tasks to all energy critical equipment to ensure that it is functioning efficiently and delivering its intended function. The details of the process are explained in this book.

ECM model is a continuous improvement maintenance regime that combines the physical preventive and predictive maintenance tasks with energy-related maintenance tasks that maintain the operational parameters of the equipment and its efficiency (i.e., motor current or fan flow rate).

ECM along with reliability centered maintenance (RCM) aims to improve the equipment reliability and energy efficiency.

The energy centered maintenance model applies to all energy critical equipment to extend its life, maintain equipment efficiency, prevent excess energy use, and reduce energy waste.

The two most important ECM strategies are:

- Preventive maintenance (time-based and usage based).
- Predictive maintenance (condition-based).

The character of the facility and equipment type determines which approaches are most effective; some combination of preventive and predictive maintenance is required to assure optimum energy efficiency.

The strategy for selecting the appropriate preventive or predictive energy centered maintenance approach involves the following:

- Consider the variety of deficiencies that may develop in equipment.
- If preventive maintenance is not adequate to detect these deficiencies, use predictive maintenance. One or a combination of several predictive technologies may be required.
- If predictive maintenance does not adequately apply or is cost prohibitive, use preventive maintenance. Develop inspection tasks to reveal the deficiencies not adequately covered by predictive maintenance.
- Decide the combination of approaches (predictive/proactive) and then determine the frequency of the particular task.

ECM provides the basis for identifying multiple low or no cost operation and maintenance practices that reduce the use of energy and enhance the efficiency of the equipment. ECM works on the concept of returning the equipment to its original operational parameters (as originally commissioned) which result in improving its energy efficiency and reduce its energy use.

Practical maintenance activities and measures that require very minimal cost or that may not require any investment can be useful measures of the energy centered maintenance program and the operation and maintenance (O&M) regime. Energy reduction can be achieved in numerous items of equipment which results in the overall decrease of building's energy consumption.

Table 3.1 illustrates cost advantages of the two different maintenance approaches in the energy centered maintenance model.

Different studies and experiments have shown that the energy consumption of equipment can be optimized and reduced by 5%–15% without a significant investment (PECI 1999). Acknowledging that significant energy reduction can be achieved by related performance operation

Table 3.1 Cost advantage of PM and PDM.

Maintenance type	Frequency	Risk of deficiency	Cost of maintenance	Cost of parts
Preventive maintenance	Time-based or operating cycles	Low – scheduled work	Low – minimum shutdown costs resulting from time-based frequency	Low – controlled and scheduled frequencies permit fewer inventory parts
Predictive maintenance	Condition-based	Lowest – deficiency avoided through detection	Lowest – condition is determined while asset is operating	Lowest – lead time permits planning of material requirements

and maintenance measures is the main reason why the energy centered maintenance model has been developed.

Energy centered maintenance model is a unique maintenance program that focuses on energy-related equipment such as air handling units, electrical motors, pumps, etc. Those types of equipment should be identified based on their energy consumption, and information should be collected from the testing and commissioning data (T&C) to compare the equipment's current behavior with its original commissioning data.

Well-functioning equipment is safe for the operation of the facility. When equipment is properly maintained, its efficiency remains high, and its operating performance remains acceptable across its lifespan, thereby minimizing the risk of any deficiencies that could increase the energy consumption of the building.

The proper execution of energy centered maintenance model should be delivered via proper job plans that are defined by the equipment's preventive and predictive maintenance needs. Equipment efficiency is significantly increased when tasks are accomplished by standard proactive programs. Job plans should be designed for each piece of equipment based on relevant factors such as the maintenance tasks, frequency, tools, duration, and craft type. The cycle of developing and implementing the job plans is the primary step in an energy centered maintenance process.

The job plans should be determined, scheduled, and updated on the organization's existing CMMS system and existing maintenance reliability program to allow a proper scheduling and monitoring. However, it is not necessary that the ECM program frequency should follow the regular PM program for cost effectiveness.

The energy centered maintenance model is a process-based methodology used to analyze and continuously improve assets and equipment maintenance and energy efficiency. This approach ensures that informed energy efficiency measures are made that will significantly impact the energy efficiency of the facility in a cost-effective maintenance model.

The model is a seven-step process that involves identifying the type of equipment and measuring its current performance and improving it (Figure 3.2); the steps are:

Step 1.0: Identify energy critical equipment.
Step 2.0: Collect data and define baseline performance.
Step 3.1: Identify energy centered maintenance inspection and frequency.
Step 3.2: Identify craft, tool, and duration.
Step 4.0: Measure equipment's current performance and compare to baseline.

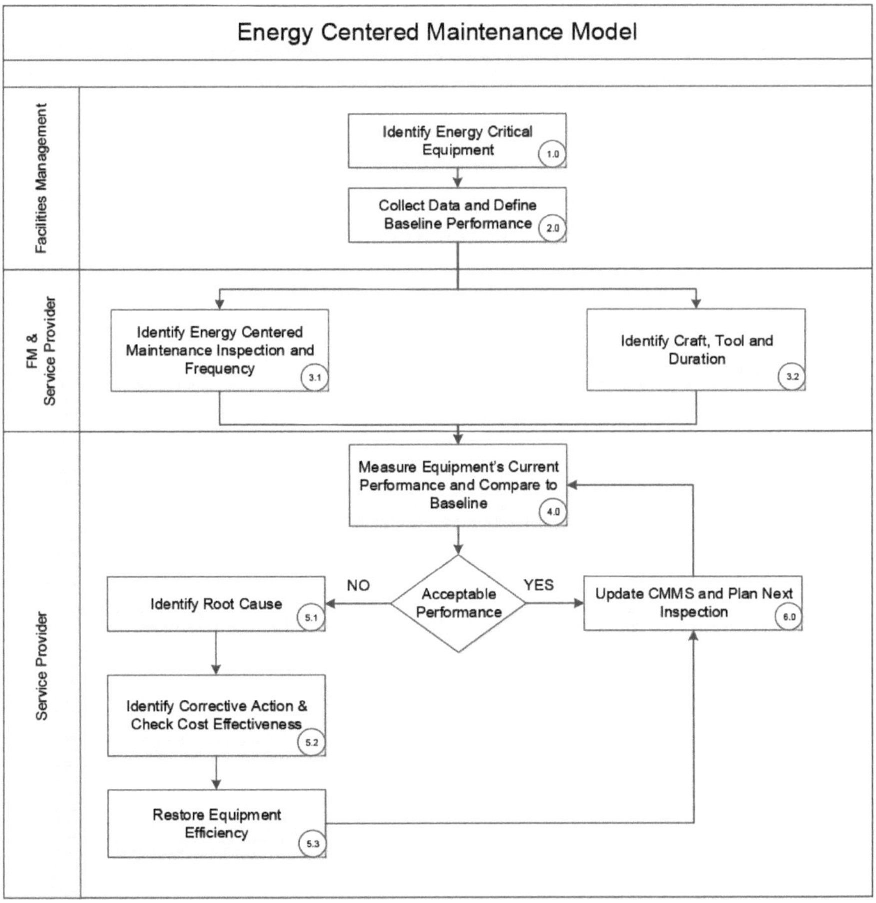

Figure 3.2 ECM model.

Step 5.1: Identify root cause.
Step 5.2: Identify corrective action and check cost effectiveness.
Step 5.3: Restore equipment efficiency.
Step 6.0: Update CMMS and plan next inspection.

3.3 Objectives of ECM

The purpose of energy centered maintenance model (ECM) is to develop a proactive maintenance approach focused on the analysis of equipment operational parameters to ensure that it is functioning by its design intent. The model will help to reduce the energy consumption of the equipment and improve its efficiency.

The ECM process's objective is to increase the energy efficiency of the equipment in a cost-effective manner using proven maintenance assessments and identifying maintenance related tasks that measure and improve the current operational behavior of the equipment.

Unlike reliability maintenance, the energy centered maintenance model does not intend to enhance equipment reliability or to prevent failures; the aim is to create maintenance tasks that prevent energy waste during equipment operation and to ensure it is delivering the intended function

Development and implementation of ECM model also have the following objectives:

- To provide educational practice of how energy consumption is related to maintenance.
- Improving maintenance regime to focus on the operational condition of the equipment.
- To identify any change in the equipment performance compared to testing and commissioning data.
- To identify improvements which can be made to increase equipment's efficiency.
- Optimizing the energy consumption of the equipment during its operation.
- Increase energy efficiency of the equipment through a low operation and maintenance cost.
- Ensure that equipment delivers the expected operational parameters as per the design intent.
- Reduce energy consumption of the facility.
- Reducing greenhouse gas emissions, mainly CO_2 emission and reducing carbon footprint, caused by energy consumption.

4

ECM Process – Equipment Identification

4.1 Step 1: Equipment Identification

The first step in developing an energy centered maintenance strategy is the identification of all energy-related systems (i.e., mechanical system) and all energy critical equipment (i.e., pumps) that should be included in the maintenance strategy of the facility.

The energy-related systems and their components differ from one building to another depending on the type of the building; for example, escalators might be available in malls but not necessarily in residential buildings.

Equipment identification defines the energy critical equipment that should be considered for energy centered maintenance approach. Energy criticality is determined based on two main factors:

- Building systems.
- Equipment energy classification code.

Those factors are focusing on identifying high energy consumption systems as well as main energy consuming equipment. Each equipment should be assigned with energy classification code based on the amount of energy consumed by the equipment and its operational hours, which in its turn decides if the equipment is feasible to be included in the energy centered maintenance model or not.

4.2 List of Energy-Related Systems

The first step is to list down all the systems and mechanical or electrical equipment that are energy related within the facility.

The following list of systems and components focus on energy centered maintenance requirements, and the equipment which is part of those systems are critical for the energy consumption of the facility.

4.2.1 Mechanical Systems

- Heating, ventilation, and air conditioning system:
  - Air handling units.
  - Fan coil units.
  - Energy recovery units.
  - Boilers.
  - Pumps.
  - Close control units.
  - Fans.
  - Cooling towers.
  - Air cooled chiller.
  - Water cooled chillers.
  - Heat exchanger.
  - Direct expansion air conditioning units.
- Water supply system:
  - Pumps.
  - Heat exchangers.
  - PRV stations.
  - Boilers.
- Drainage system:
  - Sump pumps (sewage).
- Storm water management system:
  - Rainwater pumps.
- Building transportation system:
  - Elevators.
  - Travelators.
  - Escalators.

4.2.2 Fire Fighting Systems

- Fire pumps.

4.2.3 Electrical Systems

- Motor control centers.
- Variable frequency drive (VFD).

4.2.4 Building Management System

- Two-way/three-way valve functionality.

- Differential pressure switch – DPS.
- Differential pressure transmitter – DPT.
- Airflow meters.
- Velocity meters.
- On-coil temperature and humidity sensors.
- Off-coil temperature and humidity sensors.
- Space temperature and humidity sensors.
- Thermostat functionality.
- Control logic for all equipment.

Each facility shall identify all systems and equipment applicable to their building, and the technical information about the equipment can be found in asset registers, equipment schedules, O&M manuals, as-built drawings, etc. This information will be used to define the baseline performance of the energy-related equipment and to design a maintenance checklist and plan as part of the preventive and predictive energy centered maintenance strategy.

4.3 Energy Classification Code

Energy classification is defined as the process of identifying the equipment's energy impact and criticality to the facility operation. In-depth analysis must be performed for significantly energy consuming equipment before starting energy centered maintenance model and efficiency improvement. Each system should be examined and classified by its impact on the facility operation and consumption, taking into consideration its impact on customer comfort, regulatory requirements, consumption level, and environmental impact.

The energy classification code (ECC) reflects the criticality of equipment on building energy use; a high energy classification code indicates high energy consumption equipment. The importance of assigning an energy classification code for each equipment is essential to identify which equipment will be part of the energy centered maintenance model (Table 4.1).

The ECC has a scale of 1–5, where 5 is high energy critical equipment, and 1 is low energy critical equipment. Thus, the equipment which results with ECC of 5, 4, or 3 shall be part of energy centered maintenance model, while equipment with ECC of 1 and 2 are considered of low energy impact, and, hence, it is not part of the maintenance strategy. Equipment with ECC codes 5, 4, and 3 are the ones that will be part of energy centered maintenance plan, and detailed data about the equipment should be collected to start the maintenance planning.

Table 4.1 Energy classification code.

	Energy classification	Energy impact	Description	Examples
Energy classification code	High 4 or 5	Large impact energy users	Systems with the following characteristics must be considered highly critical: • Operating profile: continuous running. • Energy capacity: high capacity.	Chilled water pumps, AHU motors, lighting, etc. • Efficiency loss on these systems will result in high associated energy costs. • Must be continuously running.
	Medium 3	Medium impact energy users	Systems with the following characteristics must be considered critical: • Operating profile: continuous running. • Energy capacity: low capacity.	Toilet exhaust fans, lighting, etc. • Efficiency loss on these systems will have a medium impact on facility energy cost. • Might be running on time schedule.
	Low 1 and 2	Low impact energy users	Systems with the following characteristics must be considered non-critical: • Operating profile: non-continuous running. • Energy capacity: vary in capacity.	Fire pumps, staircase pressurization fans, emergency lighting. • Might be high or low energy consumers. • Operates only in case of emergency.

To determine what ECC should be given for certain equipment, mathematical calculations should be made based on two measurable parameters, which are the operating profiles (run time and continuity of equipment operation in hours/day) and energy capacity (high/low) based on the connected load of the equipment.

The following steps explain the process of assigning an energy classification code for building equipment by using the following mathematical equation, graphical scale, and ECC table:

1. List down all energy-related equipment that is serving the facility.
2. Identify connected electrical load for each equipment (kW).
3. Determine total daily working hours for each equipment.
4. Calculate daily operating load (electrical consumption) of each equipment (kW.hr).
5. Calculate daily operating load (electrical consumption) of all equipment (summation of kW.hr).
6. Calculate each equipment operating weight percentage to total equipment operating load using the following equation (%).
7. Assign energy classification code for each equipment based calculated equipment operating load percentage by using the following scale.

$$\text{Single Equipment Load Percentage}(\%) = \frac{Quantity \times Operating\ Hours\left(\frac{Hours}{Day}\right) \times Connected\ Power(kW)}{Full\ Daily\ Load\ of\ All\ Equipment\left(\frac{kWh}{Day}\right)}$$

Example of calculating and assigning energy classification code:
Consider a facility that is served by multiple mechanical types of equipment, and the maintenance personnel and energy managers should be able to calculate the energy classification code for each piece of equipment to decide whether this equipment should be part of energy centered maintenance model or not.

Energy classficaiton code can be used to identify the enegy classification scale from figure 4.1

The facility is serviced by the following mechanical equipment:

- Three chilled water pumps.

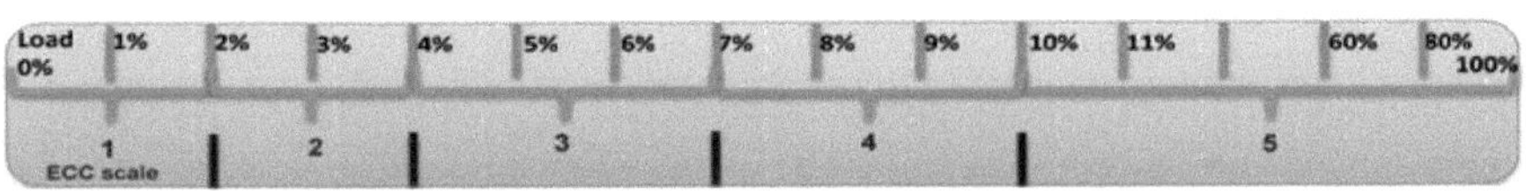

Figure 4.1 Energy classification code scale.

- Three air handling units.
- 25 fan coil units.
- Three elevators.
- Three kitchen extract fans.
- Three smoke extract fans.
- Six toilets extract fans.
- Two ejector pumps.
- One firefighting pump.

Following the steps described above, the energy managers and maintenance personnel should define the connected load for each piece of equipment (kW) and the total daily operating hours (hr) so that they can calculate the operating load (kW.hr) as described in table 4.2.

The previous table shows that the total operating load for all equipment is 9154 kW.hr per day which is required to calculate single equipment load percentage to assign the ECC code by using the scale in figure 4.2. For example, for chilled water pumps, the parameters are:

- Connected load for each pump is 50.0 kW.
- Operating hours per day for each pump is 24 hrs.
- Total operating load for all equipment is 9154 kW.hr per day.

Table 4.2 List of equipment and its related electrical consumption.

Equipment type	**Quantity**	**Connected load kW**	**Operating hours per day (hr)**	**Operating load, consumption for each equipment (kW.hr)**	**Operating load, consumption for all equipment (kW.hr)**
Chilled water pumps	3	50	24	1200	3600
Air handling units	3	15	18	270	810
Fan coil units	25	1.5	18	27	675
Elevators	3	75	12	900	2700
Kitchen extract fans	3	15	16	240	720
Kitchen supply fan	3	10	16	160	480
Smoke extract fans	3	30	0.5	15	45
Ejector pumps	2	2	16	32	64
Firefighting pump	1	120	0.5	60	60
Total operating load (electrical consumption for all equipment)					**9154**

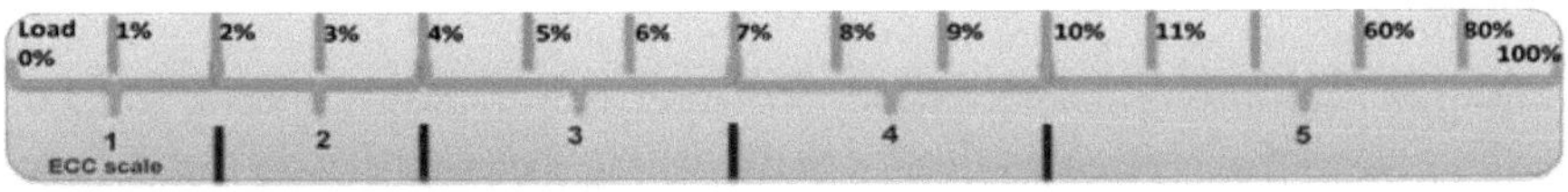

Figure 4.2 Energy classification code scale – Example 1.

$$\text{Single Equipment Load Percentage}(\%) = \frac{Quantity \times Operating\ Hours\left(\frac{Hours}{Day}\right) \times Connected\ Power(kW)}{Full\ Daily\ Load\ of\ All\ Equipment\left(\frac{kWh}{Day}\right)}$$

$$\text{Single Equipment Load Percentage}(\%) = \frac{3 \times 24\left(\frac{Hours}{Day}\right) \times 50.0\ (kW)}{9154\left(\frac{kWh}{Day}\right)}$$

Considering chilled water pump load percentage is 39.33%, and by refereeing to the scale, the energy classification code that should be assigned to the pump is 5.

Considering the same calculation criteria for each equipment type, the ECC for all equipment are as follows.

As it can be seen in table 4.3, only equipment with ECC code of 3, 4, and 5 should be part of the energy centered maintenance model. Equipment with ECC of 1 and 2 do not contribute much to the total energy consumption of the facility and may be not cost-effective to include it in the model.

Table 4.3 Calculated energy classification code.

Equipment type	Operating load, consumption for each equipment (kW.hr)	Operating load, consumption for all equipment (kW.hr)	Load % for each equipment	Energy classification code
Chilled water pumps	1200	3600	39.33%	5
Air handling units	270	810	8.85%	4
Fan coil units	27	675	7.37%	4
Elevators	900	2700	29.50%	5
Kitchen extract fans	240	720	7.87%	4
Kitchen supply fan	160	480	5.24%	3
Smoke extract fans	15	45	0.49%	2
Ejector pumps	32	64	0.70%	1
Firefighting pump	60	60	0.66%	1

Firefighting pumps remain an ideal example of low energy impact equipment, although their connected load is very high (120 kW in this example), but it works only in the unlikely case of fire. Therefore, it has no significance on the energy consumption of the facility.

5

ECM Process – Data Collection

5.1 Step 2: Data Collection and Equipment Operational Baseline

All equipment assigned with ECC codes of 5, 4, and 3 should go through data collection process which must be performed before proceeding further. Data collection is a process of gathering and measuring the required variables and information to support the energy maintenance tasks in the energy centered maintenance model.

Data collection is necessary to define the baseline for the equipment operational parameters; it focuses on obtaining the information about the design of the equipment and the testing and commissioning parameters which define the base operational behavior of the equipment.

Data collection process is needed as it ensures that data gathered, both determined and accurate, and that subsequent decisions based on the findings are valid. The process provides both a baseline from which to measure and, in particular, cases a target on what to improve.

5.2 Types of Data

The energy centered maintenance model focuses on the operational parameters of the equipment. Thus, information about equipment's profile and efficiency are essential for the execution of the model.

The type of data that need to be collected differs based on equipment type; data can be found in the testing and commissioning records, O&M manuals, as-built drawings, etc.

The following list of data is a sample of what needs to be collected, however, on a case-by-case basis; the FM team shall design this part based on the type of operating systems in their facilities.

5.2.1 Heating, Ventilation, and Air Conditioning System

- Air handling units:
  - Fan airflow rate (m^3/hr).
  - Motor power (Amps, Voltage).
  - Cooling coil pressure drop (kPa).
  - Cooling coil performance.
  - VFD data.
  - Control valve control logic.
  - On-coil/off-coil temperatures (°C).
  - Chilled water delta T (°C).
- Fan coil units:
  - Fan airflow rate (m^3/hr).
  - Motor power (Amps, Voltage).
  - Cooling coil pressure drop (kPa).
  - Cooling coil performance.
  - VFD data.
  - Control valve control logic.
  - On-coil/off-coil temperatures (°C).
  - Chilled water delta T (°C).
- Energy recovery units:
  - Fan airflow rate (m^3/hr).
  - Motor power (Amps, Voltage).
  - Effectiveness (on/off air temperature).
  - Cooling coil performance (pressure drop, temperature) – where applicable.
- Boilers:
  - Boiler efficiency.
  - Electric heater power (kW) – for electric powered boilers.
  - Fuel consumption efficiency – for fuel powered boilers.
  - In/out water characteristics.
- Pumps:
  - Pump water flow rate (lps).
  - Motor power (Amps, Voltage).
  - Pump head (kPa).
- Close control units:
  - Fan airflow rate (m^3/hr).
  - Motor power (Amps, Voltage).
  - Cooling coil pressure drop (kPa).
  - VFD data

- On-coil/off-coil temperatures (°C).
- Chilled water delta T (°C).
- Compressor power (Amps, Voltage) – for DX unit.
- Energy efficiency ratio – for DX unit.

- Fans:
  - Fan airflow rate (m^3/hr).
  - Motor power (Amps, Voltage).
  - VFD data.
- Cooling towers:
  - Water temperature in/out (°C).
  - Fan airflow rate (m^3/hr).
  - Motor power (Amps, Voltage).
  - VFD data.
- Air cooled chiller:
  - Compressor power (Amps, Voltage).
  - Energy efficiency ratio.
  - Chilled water (pressure drop).
  - Chilled water temperature in/out (°C).
  - Chilled water delta T (°C).
  - Refrigerant charge – pressure.
  - Condenser fan – motor power (Amps, Voltage).
  - VFD data.
- Heat exchanger:
  - Water temperature in/out (°C) – hot and cold sides.
  - Water pressure drop (kPa) – hot and cold sides.
- Water cooled chillers:
  - Compressor power (Amps, Voltage).
  - Energy efficiency ratio.
  - Chilled water (pressure drop).
  - Condenser (pressure drop).
  - Chilled water temperature in/out (°C).
  - Chilled water delta T (°C).
  - Refrigerant charge – pressure.
  - VFD data.
- Direct expansion air conditioning units:
  - Compressor power (Amps, Voltage).
  - Energy efficiency ratio.
- Economizers:
  - Mixing airflow rate (m^3/hr).
  - Mixing outlet airflow temperature (°C).

- Air compressors:
  - Air pressure (PSI).

5.2.2 Water Supply System

- Pumps:
  - Pump water flow rate (lps).
  - Motor power (Amps, Voltage).
  - Pump head (kPa).
- Heat exchangers:
  - Water temperature in/out (°C) – hot and cold sides.
  - Water pressure drop (kPa) – hot and cold sides.
- PRV stations:
  - On PRV pressure (Bar).
  - Off PRV pressure (Bar).
- Boilers:
  - Boiler efficiency.
  - Electric heater power (kW) – for electric powered boilers.
  - Fuel consumption efficiency – for fuel powered boilers.
  - Water temperature (°C).

5.2.3 Drainage System

- Sump pumps (sewage):
  - Pump water flow rate (lps).
  - Motor power (Amps, Voltage).
  - Pump head (kPa).
  - VFD data.

5.2.4 Storm Water Management System

- Rainwater pumps:
  - Pump water flow rate (lps).
  - Motor power (Amps, Voltage).
  - Pump head (kPa).
  - VFD data.

5.2.5 Building Transportation System

- Elevators.
- Travelators.

- Escalators:
  - Motor power (Amps, Voltage).

5.2.6 Fire Fighting Systems

- Fire pumps:
  - Pump water flow rate (lps).
  - Motor power (Amps, Voltage).
  - Pump head (kPa).
  - VFD data.

5.2.7 Electrical Systems

- Motor Control Centers:
  - In/out voltage at constant frequency (V).
  - In/out current at constant frequency (Amps).
  - Display frequency and actual out-voltage.
  - Ambient temperature.
  - Operating temperature at constant frequency.
- Variable frequency drive (VFD):
  - In/out voltage at constant frequency (V).
  - In/out current at constant frequency (Amps).
  - Display frequency and actual out-voltage.
  - Ambient temperature.
  - Operating temperature at constant frequency.

5.2.8 Lighting

- Lighting lux level.

5.2.9 Building Management System

- Two-way/three-way valve functionality:
  - Control logic.
- Differential pressure switch – DPS:
  - DPS set value.
  - Actual pressure measurement.
  - Control logic.
- Differential pressure transmitter – DPT:
  - DPT set value.
  - Actual pressure measurement.
  - Control logic.

- Airflow/velocity meters:
  - Air velocity (m^3/hr).
  - Control logic.
- On-coil temperature and humidity sensors:
  - Sensor ohmic value and actual temperature value.
- Off-coil temperature and humidity sensors:
  - Sensor ohmic value and actual humidity value.
- Space temperature and humidity sensors:
  - Sensor ohmic value and actual humidity and temperature values.
- Thermostat functionality:
  - Design set point.
  - Control logic.

5.3 Sources of Data

Data collection and information required for the original equipment design is the baseline for comparing the current operational behavior of the equipment with the intended design function. Therefore, the list of data mentioned in the previous section should be collected from the following sources:

- Drawings: construction drawings, and equipment specifications.
- Manufacturer O&M manuals: records of all manufacturer's operation and maintenance instructions.
- Testing and commissioning records: all testing and commissioning reports including readings about operational parameters of the equipment.
- Maintenance records: records of maintenance conducted during operation including consumables used (for example, lubricants, fuel, and filters).
- Procedures: all operating or maintenance procedures identified for each equipment.
- Design information: All design data that indicates the intended function of each equipment and designed operational parameters (for example, expected chilled water flow rate from a pump).
- Original testing and commissioning data: a complete set of T&C data which will be used as a benchmark to compare the current equipment performance to the original T&C.
- Re-commissioning data (if available).

When reviewing the above records, refer to the following guidelines:

- Validate all data.
- Verify that the equipment identification is correct.
- Verify that the equipment description is correct.
- Verify that the equipment specifications are correct.
- Verify that the operational parameters are defined.

6

ECM Process – ECM Inspections

6.1 Step 3: Identify ECM Inspections, Frequency, Craft, Tools, and Job Duration

Energy centered maintenance model is a proactive model that focuses on preventive and predictive maintenance types rather than reactive maintenance. The energy-related maintenance inspections should be combined with the regular preventive maintenance plans of any equipment. Therefore, when maintenance personnel conduct the regular periodic maintenance for equipment, they will also hold the energy centered maintenance inspections and do the required measurements accordingly if an improvement to the equipment performance can be determined.

Successful implementation of a maintenance inspection is critical to a smooth-running energy centered maintenance model within a facility and offers a high number of opportunities to improve equipment efficiency.

The frequency of conducting energy centered maintenance inspections differs based on the type of equipment and type of maintenance. It could be sufficient to perform the energy centered maintenance check annually for one type of equipment and semi-annually for another, based on equipment type and its energy classification code.

Energy centered maintenance inspections are defined for all equipment listed in the previous section. The frequency of the inspection to be conducted will also be specified based on the available reliability maintenance records, available information about equipment performance and efficiency, and equipment's energy classification code.

This section represents the core maintenance practice in the energy centered maintenance model. For specific items of equipment, it is required to collect the data that are related to the operational behavior of the equipment, and it is, therefore, essential that the maintenance staff should be familiar with the equipment O&M requirements to comply with the maintenance inspections specified therein.

The equipment's energy centered maintenance inspections should be carried out by the appropriately qualified staff who are capable of conducting the inspections and recording the measurements accurately and can make the right judgment about each equipment behavior.

6.2 Maintenance Records

Energy centered maintenance inspections and frequency are influenced by the reliability maintenance history of the equipment. The records are necessary to define what maintenance tasks should be conducted as part of preventive and predictive maintenance regimes once the inspection is completed and the current operational performance of the equipment is measured.

A frequently failed asset in maintaining its operational efficiency may require a different energy centered maintenance strategy to be applied than other assets, even if it is performed for the same type (i.e., two motors serving two different air handling units). Or it may be not operationally feasible or is too expensive to maintain. Hence, it should be replaced. This type of decision can be judged based on the available maintenance records of each particular asset or equipment within the facility and based on the cost effectiveness of the remedies of a certain deficiency.

Another important point is the examination of the asset over its life cycle; the records of previous maintenance contribute in calculating the total cost of ownership of the equipment since installation date.

Maintenance records should include:

- A coherent equipment repair history.
- A record of maintenance performed on equipment.
- The cost of maintenance.
- The cost of energy.
- Replacement information.
- Modification information.
- Spare parts replacement.
- Diagnostic monitoring data (if available in BMS).
- Condition assessment.
- Energy efficiency records.
- Retro-commissioning records.

Maintenance records can be used for activities such as energy efficiency analysis, energy centered maintenance inspections, preventive maintenance tasking, predictive maintenance tasking, frequency planning, and life cycle analysis.

Equipment repair history data is essential to support maintenance activities, upgrade maintenance programs, optimize equipment performance, improve equipment efficiency, plan corrective maintenance, track equipment modifications, and develop equipment and system life cycle plans.

6.3 Energy Centered Maintenance Inspections

Identifying energy centered maintenance inspection is the primary process in this guideline; the inspection should be properly defined, assigned, and sequenced for each type of equipment.

Energy centered maintenance checks should not be carried out separately from the regular reliability maintenance but should be combined with the existing preventive and predictive maintenance plans for the equipment.

The aim of this model is to amend the current reliability maintenance job plans to include those inspections that are related to the operational parameter of the equipment (i.e., efficiency). ECM inspections are defined at the end of this section for the following system's equipment:

1. Heating, ventilation, and air conditioning system.
2. Water supply system.
3. Drainage system.
4. Storm water management system.
5. Building transportation system.
6. Fire-fighting system.
7. Electrical system.
8. Building management system.

Specifying energy centered maintenance inspections should be established based on clear targets; for example, in an air handling unit, the target is to ensure that the AHU is capable of delivering the required airflow rate as intended in the design stage. This goal sets what maintenance inspections should be conducted.

Energy centered maintenance inspections help to define what elements of maintenance are required for specific equipment; an outage inspection should be specified for testing and should provide management with information necessary to control equipment performance.

The inspections should be prepared considering the following:

- Determine which deficiencies may have an impact on customer satisfaction to correct.

- Determine which deficiencies are the most cost beneficial to correct.
- Determine which deficiencies are the most performance adequate to correct.
- Determine which performance parameters are critical for equipment operation to correct.

Scheduling energy centered maintenance inspections should be performed in such a way that energy centered maintenance tasks are conducted in the proper sequence, efficiently, and within prescribed time limits.

6.4 Energy Centered Maintenance Inspection Frequency

Preventive maintenance is a time-based maintenance that should be performed with predetermined plans and frequencies and with job plans specifying how often inspection of equipment should take place. Predictive maintenance is a condition-based maintenance where any reduction in equipment efficiency (i.e., supplied flow rate) can be captured at an early stage before a deficiency in equipment performance occurs. The frequency of energy centered maintenance inspections could vary based on different parameters such as:

- Equipment operating life.
- Physical condition.
- Failure interval and failure rate.

The more frequent the maintenance inspections take place, the higher the cost but the greater the chances of maintaining equipment efficiency. Conversely, the less frequent the inspection, the less the cost but the higher the chances of increased energy use and increased energy waste intervals which result in high corrective maintenance cost. A balance between energy centered maintenance inspection, frequency, cost, and equipment efficiency should be assessed while defining the optimum ECM frequency.

In reliability maintenance, the frequency of performing preventive maintenance tasks can be calculated based on the probability of failure of a machine or based on failure rate and failure intervals, but this is for those failures which stop the equipment from performing its intended function. Energy centered maintenance is not related to this kind of failures. ECM is focused on the efficiency of the equipment while it is working; therefore, calculating specific frequencies to conduct ECM inspections is critical for cost effectiveness.

The energy centered maintenance strategy calls to perform ECM inspections as part of equipment's regular preventive maintenance job plans; it will be either based on the following:

- Annual basis:
  Where ECM inspections will be part of the annual PPM plans.
- Semi-annual basis:
  Where ECM inspection will be part of the semi-annual PPM plans.

The operation and maintenance team can revisit the energy centered maintenance inspection frequencies provided within this model depending on the actual condition of the equipment. An increase or decrease in maintenance frequencies should be analyzed based on justifications such as condition monitoring. For example, when there is no loss in equipment efficiency (i.e., pump flow rate), the energy centered maintenance frequency may be reduced.

6.5 Energy Centered Maintenance Craft, Tool, and Duration

Proper planning for the required level of craft personnel should be determined as part of the job plan. The productivity of the works done by the maintenance personnel can be optimized if the job plans specify what kind of tools are required while performing the ECM inspections and what the time duration needed to complete it is.

This section sets the criteria for selecting the right personnel, tools, and time duration required to perform energy centered maintenance inspections.

6.5.1 Craft Personnel

Energy centered maintenance inspections and job plans should be performed by a team of appropriately qualified and experienced personnel to achieve safe and efficient maintenance operations.

The facilities management team should provide the administrative and functional structure that determines the skills needed, appropriate assignments, and the performance standards for each craft group. Maintenance personnel must be skilled inefficient troubleshooting of equipment problems and must be acquainted with plant policies, procedures, systems, and equipment changes that affect their activities. Crafts personnel are usually very knowledgeable about the equipment that they maintain and understand the needs to do a quality job. Table 6.1 provides a description about the craft's personnel for different functions within building's operation

Maintenance professionals must consider all of the various requirements for scheduled and unscheduled maintenance; then compile an initial projection of the personnel requirements for the future maintenance activities. Maintenance and operations personnel are intimately familiar with the facility, systems, and equipment. Therefore, they should communicate any scheduling or task deficiencies to supervisors.

Experienced maintenance personnel should meet the following criteria:

- Understand general facility systems and equipment layout.
- Comprehend the purpose and importance of the facility's systems and equipment.
- Understand the effect of ECM work on the facility's systems.
- Assimilate industrial safety, including hazards associated with specific systems and equipment.
- Understand job-specific work practices.
- Comprehend maintenance policies and procedures.
- Be familiar with the personal protective equipment.
- Be capable to evaluate the performance of the equipment.

Craft Personnel Training:
Formal training programs should be implemented to develop and improve the knowledge and skills necessary to perform assigned functions and tasks associated with energy centered maintenance model. These training programs should be based on identified needs and must include provisions for the systematic evaluation of training effectiveness.

The training required for conducting ECM inspection is related to the testing and commissioning of the equipment; the maintenance professionals should define the training needed for their staff based on system needs.

An appropriate skill level requirement should be assigned to each ECM inspection according to its associated function as shown in the following table.

6.5.2 Tools and Special Equipment

Identifying tools and specialized equipment that are required to execute an efficient energy centered maintenance inspection is an essential process that should be planned during the development of ECM job plans and inspections. A controlled supply of the proper type, quality, and quantity of tools and special equipment serves to avoid delays in maintenance work activities and increase worker efficiency. Defining the right tools is essential to allow the maintenance personnel to measure the current performance

Table 6.1 Craft personnel – function description.

Function	Craft personnel	Function description
Heating, ventilation, and air conditioning	Mechanical technician, supervisor, engineer.	Responsible for the installation, maintenance, and repair of all building ventilating, heating, refrigerating, and cooling systems.
Plumbing	Mechanical technician, supervisor, engineer.	Responsible for the maintenance and repair of domestic water, steam, sewer, and other utility systems.
Fire protection	Mechanical technician, supervisor, engineer.	Responsible for the inspection, testing, maintenance, and repair of installed fire suppression systems (fire pumps).
Electrical	Electrical technician, supervisor, and engineer.	Responsible for the installation, maintenance, or repair of equipment for the generation, distribution, control, or utilization of electric energy.
Building management system	Low voltage and control technician, supervisor, engineer.	Responsible for installation and maintenance of all building management system components, communication, and control logic.

of the equipment which identifies if any energy waste is noticed and to determine if the equipment is underperforming, overperforming, or performing as intended. A sample list of tools and special equipment needed for ECM inspections are provided in table 16.2.

Effective control on ECM inspections is achieved when the appropriate tools and specialized equipment are available to the maintenance team for the timely and accurate execution of inspection. Policies and procedures must be in place that specifically describes the responsibilities and techniques for receiving, inspecting, handling, storing, retrieving, and issuing tools and equipment.

Tools and special equipment area include all tools, special condition-monitoring equipment, diagnostic and check out equipment, calibration equipment, measuring equipment, and so on. Tools and equipment in good condition that are easily obtainable are essential elements for reaching maintenance productivity and service goals. An appropriate tool requirement should be assigned to each ECM task according to its associated function as shown in the following table.

Tools and special equipment control systems should be periodically evaluated for their effectiveness. Specialized equipment and tools for measuring and testing should be calibrated and controlled. All phases of

Table 6.2 Tools and special equipment – sample list.

System	Example tool	Tool description
HVAC plumbing fire fighting	Thermometer	A device that measures a temperature gradient in space or a temperature of a liquid, gas, or air. Measured in °C or equivalent.
	Anemometer	A device that measures an air flow rate and air speed inside ducts and air terminals. In m^3/hr or equivalent.
	Manometer	A device that is used to measure the pressure of a liquid or gas and indicates the difference between two pressures. Measured in PSI or equivalent.
	Combustion Analyzer	A device that is used to measure different parameters related to combustion such as combustion air temperature, fuel to air ratio, combustion efficiency, etc.
	Ultrasonic leak detector	A device that is used to identify any leaks in a compressed air system such as boilers or a compressed refrigerant system such as in refrigeration systems.
	Flowmeter	A device used to measure the flow rate of a liquid such as a pump flow rate. Measured in m^3/hr or equivalent.
Electrical	Multi-meter	A device used to measure electric current, voltage, and resistance over several ranges of value. Measured in Watts, Ohms, or equivalent.
	Power analyzer	A device that measures electrical power characteristics of electrical equipment; it provides precise measurements of real power, power factor, harmonics, and efficiency.
	Electricity meter	A device that is used to gauge the amount of electricity consumed by a particular electrical equipment such as motors or lighting. Measured in kWh.
	Meggar tester	A device that is used to gauge the insulation resistance of an electrical cable and measured in kiloohms, megaohms, or gigaohms.
	Voltmeter	A device that is used to gauge the voltage in an electrical circuit or equipment. Measured in Volts.
	Thermal Camera	A device using infrared radiation to identify hot spots within electrical panels and electrical cables.
BMS	Ohmmeter	An electronic device to measure resistance in an electronic component or circuit.
	DDC simulator	A device used to test the functionality of equipment's operation.
	Logic simulator	Tools used to check the particular function of a logic related to a certain equipment or system.

procuring, receiving, inspecting, handling, storing, retrieving, and ensuring of tools and equipment should be controlled.

6.6 Calibration Program

The maintenance team should have a documented program for the control and calibration of test equipment and tools that ensure the availability of calibrated specialized equipment and tools.

The calibration program must assure:

- All calibration standards used by the calibrating agency are traceable to national or recognized standards.
- All measuring and testing equipment are kept in especially dedicated facilities to control storage, calibration, and issuance.
- Any critical equipment that was calibrated with out-of-tolerance test equipment is evaluated on time and re-calibrated as necessary.
- Any tool or measuring and test equipment with actual or suspected defects is marked and isolated to prevent its use.
- Calibration frequencies help maintain measuring and testing equipment accuracy and availability.
- The procedures used to calibrate measuring and testing equipment and tools include records for accountability and traceability of use.

6.7 Inspection Duration

Inspection duration specifies the expected time in conducting energy centered maintenance inspections for each equipment (i.e., 2 hrs to do ECM inspection on AHU). Time management data is necessary to quantify operational process shortfalls, calculating the cost of maintenance, and return on investment.

The main objective in specifying the timeline for completing the ECM inspection is the following:

- Optimizing planning and scheduling functions:
  Setting inspection duration provides the basis for better scheduling and planning of energy centered maintenance plans.
- Measure service provider productivity:
  Specifying the expected duration of each inspection will allow the FM team to identify the productivity of service providers by comparing the planned duration to the actual one. It will enable us to measure wrench time, non-productive time, or overtime that the technician spent doing the work.

- Identify actual cost of maintenance:
  Specifying the duration of inspections provides the basis for calculating the manpower cost while conducting the ECM inspections; cost measurement is essential for return-on-investment calculations and cost effectiveness that is assigned with ECM model.

Notes:

Duration of energy centered maintenance inspections might differ from one site to another based on maintainability and accessibility of the equipment. Therefore, the duration that is specified in this guideline is indicative. However, it should be specified in site maintenance plans.

The duration of inspection should count for the maintenance personnel productivity (i.e., time spent to reach the equipment is not counted in wrench time).

6.8 Energy Centered Maintenance Inspection Plans

This section lists down all energy centered maintenance tasks and job plan for all equipment listed in the previous chapter. Detailed job plan for each type of equipment is provided in tables 6.3 to 6.36.

6.8.1 Heating, Ventilation, and Air Conditioning System

Table 6.3 ECM inspection plan for air handling units.

Equipment type: air handling units				
Preventive maintenance				
Inspection	**Frequency**	**Tool**	**Craft**	**Duration**
Measure airflow rate (m^3/hr)	Annual	Anemometer	Mechanical technician	20–30 minutes
Check motor's full load current (Amps)	Quarterly	Multi-meter	Electrician	10–20 minutes
Measure cooling coil performance on full load (air off-coil temperature °C)	Annual	Thermometer	Mechanical technician	10–15 minutes
Measure cooling coil performance (pressure drop, PSI)	Annual	Manometer	Mechanical technician	10–15 minutes
Measure variable frequency drive effectiveness	Annual	Multi-meter	Electrician	45–60 minutes
Measure two-way/three-way control valves response to space temperature	Quarterly	DDC Simulator	Control technician/ electrician	20–30 minutes
Measure chilled water temperature difference (delta T) °C	Quarterly	Thermometer	Mechanical technician	15–20 minutes

Predictive maintenance				
Inspection	**Frequency**	**Tool**	**Craft**	**Duration**
Measure airflow rate (m³/hr)	Continues	Flowmeter (connected on BMS)	BMS operator	5–10 minutes
Measure air off-coil temperature (°C)	Continues	Thermometer (connected on BMS)	BMS operator	5–10 minutes
Measure fan's motor power consumption (kWh)	Continues	Electricity meter	BMS operator	5–10 minutes
Measure chilled water temperature difference (delta T) °C	Continues	Thermometer	Mechanical technician	5–10 minutes

Table 6.4 ECM inspection plan for fan coil units.

Equipment type: fan coil unit				
Preventive maintenance				
Inspection	**Frequency**	**Tool**	**Craft**	**Duration**
Measure airflow rate (m³/hr)	Annual	Anemometer	Mechanical technician	20–30 minutes
Check motor's full load current (Amps)	Quarterly	Multi-meter	Electrician	10–20 minutes
Measure cooling coil performance on full load (air off-coil temperature °C)	Annual	Thermometer	Mechanical technician	10–15 minutes
Measure cooling coil performance (pressure drop, PSI)	Annual	Manometer	Mechanical technician	10–15 minutes
Measure variable frequency drive effectiveness	Annual	Multi-meter	Electrician	45–60 minutes
Measure two-way/three-way control valves response to space temperature	Quarterly	DDC Simulator	Control Technician/ electrician	20–30 minutes
Measure chilled water temperature difference (delta T) °C	Quarterly	Thermometer	Mechanical technician	15–20 minutes
Predictive maintenance				
Inspection	**Frequency**	**Tool**	**Craft**	**Duration**
Measure airflow rate (m³/hr)	Continues	Flowmeter (connected on BMS)	BMS operator	5–10 minutes
Measure air off-coil temperature (°C)	Continues	Thermometer (connected on BMS)	BMS operator	5–10 minutes

Table 6.4 Continued.

Predictive maintenance				
Inspection	**Frequency**	**Tool**	**Craft**	**Duration**
Measure fan's motor power consumption (kWh)	Continues	Electricity Meter	BMS operator	5–10 minutes
Measure chilled water temperature difference (delta T) °C	Continues	Thermometer	Mechanical technician	5–10 minutes

Note: Predictive maintenance measures for fan coil units may be implemented for high energy consumption FCUs. It may not be practical to implement it on all FCUs within the facility.

Table 6.5 ECM inspection plan for energy recovery units (i.e., heat wheels).

Equipment type: energy recovery units				
Preventive maintenance				
Inspection	**Frequency**	**Tool**	**Craft**	**Duration**
Measure airflow rate (m^3/hr)	Annual	Anemometer	Mechanical technician	20–30 minutes
Check motor's full load current (Amps)	Quarterly	Multi-meter	electrician	10–20 minutes
Measure energy recovery performance on full load (air off-coil temperature °C)	Annual	Thermometer	Mechanical technician	10–15 minutes
Measure cooling coil performance (pressure drop, PSI)	Annual	Manometer	Mechanical technician	10–15 minutes
Predictive maintenance				
Inspection	**Frequency**	**Tool**	**Craft**	**Duration**
Measure airflow rate (m^3/hr)	Continues	Flowmeter (connected on BMS)	BMS operator	5–10 minutes
Measure air off-coil temperature (°C)	Continues	Thermometer (connected on BMS)	BMS operator	5–10 minutes
Measure fan's motor power consumption (kWh)	Continues	Electricity meter	BMS operator	5–10 minutes

Table 6.6 ECM inspection plan for boilers.

Equipment type: boilers				
Preventive maintenance				
Inspection	**Frequency**	**Tool**	**Craft**	**Duration**
Measure fuel combustion efficiency	Bi-annual	Combustion analyzer	Mechanical technician	30–45 minutes
Inspect steam leakage	Monthly	Ultrasonic steam leak detector	Mechanical technician	30–45 minutes
Predictive maintenance				
Inspection	**Frequency**	**Tool**	**Craft**	**Duration**
Outlet water temperature (°C)	Continues	Thermometer (connected on BMS)	BMS operator	5–10 minutes
Primary system water pressure (PSI)	Continues	Manometer (connected on BMS)	BMS operator	5–10 minutes

Table 6.7. ECM inspection plan for pumps.

Equipment type: chilled water pumps				
Preventive maintenance				
Inspection	**Frequency**	**Tool**	**Craft**	**Duration**
Measure full speed water flow rate (gpm, lps)	Annual	Anemometer	Mechanical technician	60–90 minutes
Check motor's full load current (Amps)	Quarterly	Multi-meter	Electrician	10–20 minutes
Measure variable frequency drive effectiveness	Annual	Multi-meter	Electrician	45–60 minutes
Predictive maintenance				
Inspection	**Frequency**	**Tool**	**Craft**	**Duration**
Measure water flow rate (gpm lps)	Continues	Flowmeter (connected on BMS)	BMS operator	5–10 minutes
Measure pump's motor power consumption (kWh)	Continues	Electricity meter (connected on BMS)	BMS operator	5–10 minutes
Motor running current (Amps)	Continues	Multi-meter	BMS operator	10–15 minutes

Table 6.8 ECM inspection plan for close control units.

Equipment type: close control units				
Preventive maintenance				
Inspection	**Frequency**	**Tool**	**Craft**	**Duration**
Measure airflow rate (m^3/hr)	Annual	Anemometer	Mechanical technician	20–30 minutes
Check fan motor's full load current (Amps)	Quarterly	Multi-meter	Electrician	10–20 minutes
Check DX unit compressor full load current (Amps)	Quarterly	Multi-meter	Electrician	10–20 minutes
Measure cooling coil performance on full load (air off-coil temperature °C)	Annual	Thermometer	Mechanical technician	10–15 minutes
Measure cooling coil performance (pressure drop, PSI)	Annual	Manometer	Mechanical technician	10–15 minutes
Measure variable frequency drive effectiveness	Annual	Multi-meter	Electrician	45–60 minutes
Measure two-way/three-way control valves response to space temperature	Quarterly	DDC simulator	Control technician/ electrician	20–30 minutes
Measure chilled water temperature difference (delta T) °C	Quarterly	Thermometer	Mechanical technician	15–20 minutes
Measure DX unit energy efficiency ratio (EER0	Quarterly	Multi-meter	Mechanical technician/ electrician	30–45 minutes
Predictive maintenance				
Inspection	**Frequency**	**Tool**	**Craft**	**Duration**
Measure airflow rate (m^3/hr)	Continues	Flowmeter (connected on BMS)	BMS operator	5–10 minutes
Measure air off-coil temperature (°C)	Continues	Thermometer (connected on BMS)	BMS operator	5–10 minutes
Check DX unit compressor full load current (Amps)	Continues	Electricity meter	BMS operator	5–10 minutes
Measure chilled water temperature difference (delta T) °C	Continues	Thermometer	Mechanical technician	5–10 minutes

Table 6.9 ECM inspection plan for fans.

Equipment type: fans				
Preventive maintenance				
Inspection	**Frequency**	**Tool**	**Craft**	**Duration**
Measure airflow rate (m^3/hr)	Annual	Anemometer	Mechanical technician	20–30 minutes
Check motor's full load current (Amps)	Quarterly	Multi-meter	Electrician	10–20 minutes
Measure variable frequency drive effectiveness	Annual	Multi-meter	Electrician	45–60 minutes
Predictive maintenance				
Inspection	**Frequency**	**Tool**	**Craft**	**Duration**
Measure airflow rate (m^3/hr)	Continues	Flowmeter (connected on BMS)	BMS operator	5–10 minutes
Measure pump's motor power consumption (kWh)	Continues	Electricity meter (connected on BMS)	BMS operator	5–10 minutes

Table 6.10 ECM inspection plan for cooling towers.

Equipment type: cooling towers				
Preventive maintenance				
Inspection	**Frequency**	**Tool**	**Craft**	**Duration**
Check fan motor's full load current (Amps)	Bi-annual	Multi-meter	Electrician	20–30 minutes
Cooling tower range (in-out water temperature) – full load (°C)	Monthly	Thermometer	Mechanical technician	10–20 minutes
Measure variable frequency drive effectiveness	Annual	Multi-meter	Electrician	45–60 minutes
Predictive Maintenance				
Inspection	**Frequency**	**Tool**	**Craft**	**Duration**
Check fan motor's current (Amps)	Continues	Electricity meter (connected on BMS)	BMS operator	5–10 minutes
Cooling tower range (in-out water temperature) – full load (°C)	Continues	Thermometer (connected on BMS)	BMS operator	5–10 minutes

Table 6.11 ECM inspection plan for air cooled chillers.

Equipment type: air cooled chillers				
Preventive maintenance				
Inspection	**Frequency**	**Tool**	**Craft**	**Duration**
Check compressor motor's full load current (Amps)	Bi-annual	Multi-meter	Electrician	10–20 minutes
Check condenser fan motor's full load current (Amps)	Annual	Multi-meter	Electrician	10–20 minutes
Evaporator pressure drop (PSI)	Bi-annual	Manometer	Mechanical technician	30–45 minutes
Refrigerant leaks test	Quarterly	Ultrasonic leak detector	Mechanical technician	30–45 minutes
Measure variable frequency drive effectiveness	Annual	Multi-meter	electrician	45–60 minutes
Predictive maintenance				
Inspection	**Frequency**	**Tool**	**Craft**	**Duration**
Check compressor motor's current (Amps)	Continues	Electricity meter (connected on BMS)	BMS operator	5–10 minutes
Chilled water supply temperature (°C)	Continues	Thermometer (connected on BMS)	BMS operator	5–10 minutes
Operating pressure (PSI)	Continues	Manometer (connected on BMS)	BMS operator	5–10 minutes

Table 6.12 ECM inspection plan for water cooled chillers.

Equipment type: water cooled chillers				
Preventive maintenance				
Inspection	**Frequency**	**Tool**	**Craft**	**Duration**
Check compressor motor's full load current (Amps)	Bi-annual	Multi-meter	Electrician	10–20 minutes
Evaporator pressure drop (PSI)	Bi-annual	Manometer	Mechanical technician	30–45 minutes
Condenser pressure drop (PSI)	Bi-annual	Manometer	Mechanical technician	30–45 minutes
Refrigerant leaks test	Quarterly	Ultrasonic leak detector	Mechanical technician	30–45 minutes
Measure variable frequency drive effectiveness	Annual	Multi-meter	Electrician	45–60 minutes

Predictive maintenance				
Inspection	**Frequency**	**Tool**	**Craft**	**Duration**
Check compressor motor's current (Amps)	Continues	Electricity meter (connected on BMS)	BMS operator	5–10 minutes
Chilled water supply temperature (°C)	Continues	Thermometer (connected on BMS)	BMS operator	5–10 minutes
Operating pressure (PSI)	Continues	Manometer (connected on BMS)	BMS operator	5–10 minutes

Table 6.13 ECM inspection plan for heat exchangers.

Equipment type: heat exchangers				
Preventive maintenance				
Inspection	**Frequency**	**Tool**	**Craft**	**Duration**
Measure pressure drop (PSI)	Bi-annual	Manometer	Mechanical technician	60–90 minutes
Heat exchanger effectiveness (%) – (in/out temperature)	Bi-annual	Thermometer	Mechanical technician	20–30 minutes
Predictive maintenance				
Inspection	**Frequency**	**Tool**	**Craft**	**Duration**
Chilled water supply temperature (°C)	Continues	Thermometer (connected on BMS)	BMS operator	5–10 minutes

Table 6.14 ECM inspection plan for direct expansion air conditioners (DX units).

Equipment type: DX units				
Preventive maintenance				
Inspection	**Frequency**	**Tool**	**Craft**	**Duration**
Compressor motor's full load current (Amps)	Bi-annual	Multi-meter	Electrician	10–20 minutes
Refrigerant leaks test	Quarterly	Ultrasonic leak detector	Mechanical technician	30–45 minutes
Predictive maintenance				
Inspection	**Frequency**	**Tool**	**Craft**	**Duration**
Check compressor motor's current (Amps)	Continues	Electricity meter (connected on BMS)	BMS operator	5–10 minutes

6.8.2 Economizers

Table 6.15 ECM inspection plan for economizers.

Equipment type: economizers				
Preventive maintenance				
Inspection	**Frequency**	**Tool**	**Craft**	**Duration**
Supply air flow rate after mixing (m³/hr)	Bi-annual	Anemometer	Mechanical technician	20–30 minutes
Supply air flow rate temperature after mixing °C	Bi-annual	Thermometer	Mechanical technician	20–30 minutes
	Predictive maintenance			
Inspection	**Frequency**	**Tool**	**Craft**	**Duration**
Supply air flow rate after mixing (m³/hr)	Continues	Anemometer (connected on BMS)	BMS operator	5–10 minutes
Supply air flow rate temperature after mixing (°C)	Continues	Thermometer (connected on BMS)	BMS operator	5–10 minutes

Table 6.16 ECM inspection plan for air compressors.

Equipment type: air compressors				
Preventive maintenance				
Inspection	**Frequency**	**Tool**	**Craft**	**Duration**
Produced air pressure (PSI)	Bi-annual	Manometer	Mechanical technician	20-30 minutes

6.8.3 Water Supply System

Table 6.17 ECM inspection plan for domestic water pump set, irrigation pump, and water features pumps.

Equipment type: pumps				
Preventive maintenance				
Inspection	**Frequency**	**Tool**	**Craft**	**Duration**
Measure full speed water flow rate (gpm, lps)	Annual	Anemometer	Mechanical technician	60–90 minutes
Check motor's full load current (Amps)	Quarterly	Multi-meter	Electrician	10–20 minutes
Measure variable frequency drive effectiveness	Annual	Multi-meter	Electrician	45–60 minutes

Predictive maintenance				
Inspection	**Frequency**	**Tool**	**Craft**	**Duration**
Measure water flow rate (gpm, lps)	Continues	Flowmeter (connected on BMS)	BMS operator	5–10 minutes
Measure pump's motor power consumption (kWh)	Continues	Electricity meter (connected on BMS)	BMS operator	5–10 minutes
Motor running current	Continues	Multi-meter (connected on BMS)	BMS operator	5–10 minutes

Table 6.18 ECM inspection plan for heat exchangers.

Equipment type: heat exchangers				
Preventive maintenance				
Inspection	**Frequency**	**Tool**	**Craft**	**Duration**
Measure pressure drop (PSI)	Bi-annual	Manometer	Mechanical technician	60–90 minutes
Heat exchanger effectiveness (%) – (in/out temperature)	Bi-annual	Thermometer	Mechanical technician	20–30 minutes
Predictive maintenance				
Inspection	**Frequency**	**Tool**	**Craft**	**Duration**
Chilled water supply temperature (°C)	Continues	Thermometer (BMS)	BMS operator	5–10 minutes

Table 6.19 ECM inspection plan for pressure reducing valve stations.

Equipment type: PRV stations				
Preventive maintenance				
Inspection	**Frequency**	**Tool**	**Craft**	**Duration**
Measure on/off pressure on critical PRVs	Annual	Manometer and calibration kit	Mechanical technician	30–45 minutes per PRV station

Note: Only main PRVs should be recalibrated. For example, recalibrate PRVs mainly on domestic water lines entering floors.

Table 6.20 ECM inspection plan for boilers.

Equipment type: boilers				
Preventive maintenance				
Inspection	**Frequency**	**Tool**	**Craft**	**Duration**
Measure fuel combustion efficiency	Bi-annual	Combustion analyzer	Mechanical technician	30–45 minutes
Inspect steam leakage	Monthly	Ultrasonic steam leak detector	Mechanical technician	30–45 minutes
Predictive maintenance				
Inspection	**Frequency**	**Tool**	**Craft**	**Duration**
Outlet water temperature (°C)	Continues	Thermometer (connected on BMS)	BMS operator	5–10 minutes
Primary system water pressure (PSI)	Continues	Manometer (connected on BMS)	BMS operator	5–10 minutes

6.8.4 Drainage System

Table 6.21 ECM inspection plan for sump pumps.

Equipment type: pumps				
Preventive maintenance				
Inspection	**Frequency**	**Tool**	**Craft**	**Duration**
Measure full speed water flow rate (gpm, lps)	Annual	Anemometer	Mechanical technician	60–90 minutes
Check motor's full load current (Amps)	Quarterly	Multi-meter	Electrician	10–20 minutes
Measure VFD effectiveness	Annual	Multi-meter	Electrician	45–60 minutes
Predictive maintenance				
Inspection	**Frequency**	**Tool**	**Craft**	**Duration**
Measure water flow rate (gpm, lps)	Continues	Flowmeter (on BMS)	BMS operator	5–10 minutes
Measure pump's motor power consumption (kWh)	Continues	Electricity meter (on BMS)	BMS operator	5–10 minutes
Motor running current (Amps)	Continues	Multi-meter	BMS operator	5–10 minutes

6.8.5 Storm Water Management System

Table 6.22 ECM inspection plan for rain water pumps.

Equipment type: pumps				
Preventive maintenance				
Inspection	**Frequency**	**Tool**	**Craft**	**Duration**
Measure full speed water flow rate (gpm, lps)	Annual	Anemometer	Mechanical technician	60–90 minutes
Check motor's full load current (Amps)	Quarterly	Multi-meter	Electrician	10–20 minutes
Measure VFD effectiveness	Annual	Multi-meter	Electrician	45–60 minutes
Predictive maintenance				
Inspection	**Frequency**	**Tool**	**Craft**	**Duration**
Measure water flow rate (gpm, lps)	Continues	Flowmeter (on BMS)	BMS operator	5–10 minutes
Measure pump's motor power consumption (kWh)	Continues	Electricity meter (on BMS)	BMS operator	5–10 minutes
Motor running current (Amps)	Continues	Multi-meter	BMS operator	5–10 minutes

6.8.6 Building Transportation System

Table 6.23 ECM inspection plan for travelators and escalators.

Equipment type: travelators and escalators				
Preventive maintenance				
Inspection	**Frequency**	**Tool**	**Craft**	**Duration**
Check motor's full load current (Amps)	Quarterly	Multi-meter	Electrician	20–30 minutes
Auto start/stop command	Quarterly	Multi-meter	Electrician	20–30 minutes
Predictive maintenance				
Inspection	**Frequency**	**Tool**	**Craft**	**Duration**
Measure motor power consumption (kWh)	Continues	Electricity meter (connected on BMS)	BMS operator	5–10 minutes

Table 6.24 ECM inspection plan for elevators.

Equipment type: elevators				
Preventive maintenance				
Inspection	**Frequency**	**Tool**	**Craft**	**Duration**
Check motor's full load current (Amps)	Quarterly	Multi-meter	Electrician	20–30 minutes
Predictive maintenance				
Inspection	**Frequency**	**Tool**	**Craft**	**Duration**
Measure motor power consumption (kWh)	Continues	Electricity meter (connected on BMS)	BMS operator	5–10 minutes

6.8.7 Fire Fighting System

Table 6.25 ECM inspection plan for fire-fighting pumps.

Equipment type: pumps				
Preventive maintenance				
Inspection	**Frequency**	**Tool**	**Craft**	**Duration**
Measure full speed water flow rate (gpm, lps)	Annual	Anemometer	Mechanical technician	60–90 minutes
Check motor's full load current (Amps)	Quarterly	Multi-meter	Electrician	10–20 minutes
Measure variable frequency drive effectiveness	Annual	Multi-meter	Electrician	45–60 minutes
Predictive maintenance				
Inspection	**Frequency**	**Tool**	**Craft**	**Duration**
Measure water flow rate (gpm, lps)	Continues	Flowmeter (on BMS)	BMS operator	5–10 minutes
Measure pump's motor power consumption (kWh)	Continues	Electricity meter (on BMS)	BMS operator	5–10 minutes
Motor running current (Amps)	Continues	Multi-meter	BMS operator	5–10 minutes

6.8.8 Electrical System

Table 6.26 ECM inspection plan for motor control center.

Equipment type: MCC				
Preventive maintenance				
Inspection	**Frequency**	**Tool**	**Craft**	**Duration**
In/out voltage at constant frequency	Annual	Voltmeter	Electrician	10–15 minutes
In/out current at constant frequency	Annual	Multi-meter	Electrician	10–15 minutes
Operating temperature	Quarterly	Thermal camera	Electrician	10–15 minutes
		Predictive maintenance		
Inspection	**Frequency**	**Tool**	**Craft**	**Duration**
Verify the out-voltage with frequency	Quarterly	Multi-meter	Electrician	10–15 minutes
Ambient temperature	Continues	Thermometer	Electrician	10–15 minutes

Table 6.27 ECM inspection plan for variable frequency drive.

Equipment type: VFD				
Preventive maintenance				
Inspection	**Frequency**	**Tool**	**Craft**	**Duration**
In/out voltage at constant frequency	Annual	Voltmeter	Electrician	10–15 minutes
In/out current at constant frequency	Annual	Multi-meter	Electrician	10–15 minutes
Operating temperature	Quarterly	Thermal camera	Electrician	10–15 minutes
Predictive maintenance				
Inspection	**Frequency**	**Tool**	**Craft**	**Duration**
Verify the out-voltage with frequency	Quarterly	Multi-meter	Electrician	10–15 minutes
Ambient temperature	Continues	Thermometer	Electrician	10–15 minutes

Table 6.28 ECM inspection p lan for light bulbs.

Equipment type: light bulbs				
Preventive maintenance				
Inspection	**Frequency**	**Tool**	**Craft**	**Duration**
Lux level (foot-candle)	Annual	Lux meter	Electrician	10–15 minutes per light

6.8.9 Building Management System - BMS

Table 6.29 ECM inspection plan for two-way control valve.

Equipment type: two-way control valve				
Preventive maintenance				
Inspection	**Frequency**	**Tool**	**Craft**	**Duration**
Check actuator response in line with control signal	Quarterly	Voltmeter/ visual	BMS and mechanical technicians	15–20 minutes
Check feedback correspondence to control signal	Quarterly	Voltmeter/ visual	BMS and mechanical technicians	15–20 minutes
Check response time with respect to command	Quarterly	Visual	BMS and mechanical technicians	15–20 minutes
Predictive maintenance				
Inspection	**Frequency**	**Tool**	**Craft**	**Duration**
Valves response to AHU/ FCU temperature set point	Continuous	Frontend software	BMS operator	5–10 minutes

Table 6.30 ECM inspection plan for differential pressure switch.

Equipment type: DPS				
Preventive maintenance				
Inspection	**Frequency**	**Tool**	**Craft**	**Duration**
Check DPS functionality with respect to control signal	Quarterly	Multi-meter/ visual	BMS and mechanical technicians	15–20 minutes
Conduct continuity loop testing	Quarterly	Multi-meter	BMS and mechanical technicians	15–20 minutes
Compare differential pressure value on frontend and on site	Quarterly	Calibration manometer	BMS and mechanical technicians	15–20 minutes
Predictive maintenance				
Inspection	**Frequency**	**Tool**	**Craft**	**Duration**
Differential pressure value	Continuous	Frontend software	BMS operator	5–10 minutes

Table 6.31 ECM inspection plan for differential pressure transmitter.

Equipment type: DPT				
Preventive maintenance				
Inspection	**Frequency**	**Tool**	**Craft**	**Duration**
Check DPT response with respect to control signal	Quarterly	Multi-meter/ visual	BMS and mechanical technicians	15–20 minutes
Conduct continuity loop testing	Quarterly	Multi-meter	BMS and mechanical technicians	15–20 minutes
Compare differential pressure value on frontend and on site	Quarterly	Calibration manometer	BMS and mechanical technicians	15–20 minutes
Predictive maintenance				
Inspection	**Frequency**	**Tool**	**Craft**	**Duration**
Differential pressure value	Continuous	Frontend software	BMS operator	5–10 minutes

Table 6.32 ECM inspection plan for flow rate/velocity meters.

Equipment type: flow rate/ velocity meters				
Preventive maintenance				
Inspection	**Frequency**	**Tool**	**Craft**	**Duration**
Compare flow rate value on frontend and actual on-site value	Bi-annual	Calibration flowmeter/ velocity meter	BMS and mechanical technicians	15–20 minutes
Conduct continuity loop testing	Quarterly	Multi-meter	BMS and mechanical technicians	15–20 minutes
Check flow rate response to software command (increase/decrease)	Quarterly	Multi-meter/ flowmeter	BMS and mechanical technicians	15–20 minutes
Predictive maintenance				
Inspection	**Frequency**	**Tool**	**Craft**	**Duration**
Flow rate/velocity value	Continuous	Frontend software	BMS operator	5–10 minutes

Table 6.33 ECM inspection plan for cooling coil temperature and humidity sensors.

Equipment type: cooling coil temperature and humidity sensors				
Preventive maintenance				
Inspection	**Frequency**	**Tool**	**Craft**	**Duration**
Check the sensor ohmic value and actual temperature where it is installed	Quarterly	Thermometer	BMS technician	10–15 minutes
Predictive maintenance				
Inspection	**Frequency**	**Tool**	**Craft**	**Duration**
Frontend recorded temperature value	Continuous	Frontend software	BMS operator	5–10 minutes

Table 6.34 ECM inspection plan for chilled water temperature.

Equipment type: chilled water temperature				
Preventive maintenance				
Inspection	**Frequency**	**Tool**	**Craft**	**Duration**
Check the sensor ohmic value and actual temperature where it is installed	Quarterly	Thermometer	BMS technician	10–15 minutes
Predictive maintenance				
Inspection	**Frequency**	**Tool**	**Craft**	**Duration**
Frontend recorded temperature value	Continuous	Frontend software	BMS operator	5–10 minutes

Table 6.35 ECM inspection plan for space/return air temperature and humidity sensors.

Equipment type: space/ return air temperature and humidity sensors				
Preventive maintenance				
Inspection	**Frequency**	**Tool**	**Craft**	**Duration**
Check the sensor ohmic value and actual temperature where it is installed	Quarterly	Thermometer	BMS technician	10–15 minutes
Check response of two-way valve to space/ return air temperature	Quarterly	Physical	BMS technician	10–15 minutes
Predictive maintenance				
Inspection	**Frequency**	**Tool**	**Craft**	**Duration**
Frontend recorded temperature value	Continuous	Frontend software	BMS operator	5–10 minutes

Table 6.36 ECM inspection plan for control logic for all equipment controlled by BMS.

Equipment type: control logic for all equipment				
Preventive maintenance				
Inspection	**Frequency**	**Tool**	**Craft**	**Duration**
Control logic for all equipment that are part of energy centered maintenance process.				
For example:				
AHUs	Quarterly	Logic simulator	BMS technician/ mechanical technician/ electrical technician	20–30 minutes
FCUs				15–20 minutes
FAHUs				20–30 minutes
Heat recovery units				15–20 minutes
Close control units				15–20 minutes
Pumps				60–90 minutes
Water features				60–90 minutes
Chillers				60–90 minutes
Heat exchangers				60–90 minutes
Fans				15–20 minutes
Lifts and escalators				15–20 minutes
Travelators				15–20 minutes
Boilers				60–90 minutes
Cooling towers				60–45 minutes
DX units				15–20 minutes
MCC				15–20 minutes
VFDs				20-30 minutes

7

ECM Process – Measuring Equipment Current Performance

7.1 Step 4: Measuring Equipment's Current Performance and Comparing to Baseline

The maintenance personnel should be capable of measuring the current performance of the equipment during energy centered maintenance inspection. Current performance defines the actual operational condition of the equipment which will be compared to the baseline performance as recorded in the testing and commissioning phase.

Measuring the equipment's current performance involves collecting and analyzing actual data about the operational parameters of the equipment. Parameters such as equipment's efficiency and power consumption should be measured. Those data will then help in determining if any part of the equipment is not delivering its intended function, which results in identifying what kind of corrective actions should be done to improve and restore the operational efficiency of the equipment.

7.2 Measuring Equipment's Current Performance

Measuring equipment's performance is usually performed in conjunction with the planned reliability maintenance job plans (linked with regular PPM inspections), for each energy critical assets (ECC 5, 4, and 3). The requirement is to give an indication of the expected operational condition of each equipment, indicate any possible action requirements to improve or restore the performance of the equipment, and enable better planning of future maintenance tasks.

Once the current operational condition of the equipment is conducted for ECC 5, 4, and 3 assets, a comprehensive, balanced, proactive maintenance strategy can be developed to restore and maintain the performance of the asset. This process will:

- Identify the current operational condition of the assets which will be compared to the baseline data collected about each equipment.
- Help to identify the cause of performance deterioration and corrective action.
- Calculate the cost of repair and the cost effectiveness.
- Improving the level of maintenance.

Measuring current equipment's performance is a purposeful examination of the equipment with the intent to benchmark to baseline information or identify problems. The key to the success lies in the dedication of time for the particular purpose of assessment and the qualifications of the inspector. No intention is made during the operational assessment to repair the identified problems immediately; the workforce merely recognizes them.

The measurements shall be compared with the baseline value, and the maintenance personnel shall be capable of judging if the equipment is overperforming, underperforming, or performing as intended. For example, measuring motor running current may be acceptable if it is running within an acceptable range compared to testing and commissioning values or an AHU is delivering an acceptable range of airflow compared to original data.

The principal of regular measurement of equipment efficiency aim of the balanced, proactive maintenance approach is to uncover deficiencies or energy waste before they are on early stage to reduce its impact on the overall energy consumption of the facility. The sooner the problems are found, the greater the opportunity for planning, gathering materials, coordinating outages, estimating, allocating resources, and restoring equipment performance.

A sample equipment list, with defined performance measures (i.e., motor consumption) that should be measured, is defined in Chapter 9. Similarly, acceptable performance range is also determined; thus, if the equipment is performing within this range, then the equipment is considered as performing as intended. Otherwise, a root-cause analysis should be conducted.

7.3 Root-Cause Analysis

The aim of root-cause analysis process is to prevent the reoccurrence of a functional failure in reliability maintenance regime. Similarly, in energy centered maintenance, the goal is to eliminate operational deficiencies of the equipment and to reduce its energy waste. The cause of the energy waste or functional deficiency must be determined through a thorough investigation. This investigation would include researching equipment history, evaluating current conditions and following RCA methods.

Energy centered maintenance model recommends that all operational deficiencies and recognized energy wastes from all energy critical equipment are investigated to eliminate it or reduce it. Root-cause analysis can assist in minimizing or eliminating energy waste during equipment operation by developing effective corrective actions and by adjusting the balanced proactive maintenance strategy approach to best suit the equipment.

The RCA process is applied uniformly to determine the cause of an operational deficiency or energy waste. This process is established on evidence-based causal thinking. RCA can be done by one person troubleshooting a problem or by a team of people investigating a major event. It is the same process throughout the organization.

There are two types of problems:

- **Rule-based** problems that have one solution.
- **Event-based** problems that have many possible solutions.

Most problems that occur in a facility are event-based problems, i.e., there are many possible solutions.

RCA is an organized methodology or process for determining the cause and effect of a particular failure (or operational deficiency/energy waste) and finding solutions to minimize or eliminate them from reoccurrence. The RCA process plays an important role in:

- Minimizing or eliminating costly failures or operational deficiencies.
- Developing effective corrective actions.
- Creating a prevention culture based on cause and effect.
- Focusing efforts on effective solutions.

The process of root-cause analysis involves the following two steps.

7.3.1 Define the Problem

The process starts by using the information collected during the inspection which identified certain problems such as low equipment performance, low energy efficiency, and high energy waste. The problem should then be further investigated by defining the following four main elements:

1. **What**: what the actual problem is. This may be the indicator or symptom of the problem. It is possible that there is more than one problem for any given event or incident.

2. **When**: the actual time the problem occurred and the status of the system or process when the problem occurred.
3. **Where**: where the problem occurred. This may be the site, facility, building, system, component, or the equipment ID.
4. **Significance**: the significance of the problem. The significance relates to the efficiency and energy consumption of the equipment.

7.3.2 Use Cause and Effect Chart

Cause and effect chart (so-called fishbone diagram or Ishikawa diagram) is a systematic method for identifying all possible causes for a certain effect or a problem. The diagram helps in suggesting all possible causes about energy wastes or operational deficiencies that are found in equipment during energy centered maintenance inspection. The diagram was created by Karou Ishikawa (1968) (Figure 7.1).

Once the problem is defined, a cause and effect chart should be prepared by identifying the primary cause of the problem. The causes should then be evaluated by asking "Why" the cause could happen. The process should continue to identify all possible causes of the problem; those causes should then have enough evidence that it is related to the equipment, and the causes will then be used to define the required corrective actions and to calculate cost effectiveness to solve the problem.

The motivation for using cause and effect chart is to understand deficiency data and reduce costly energy wastes by appropriate, cost-effective

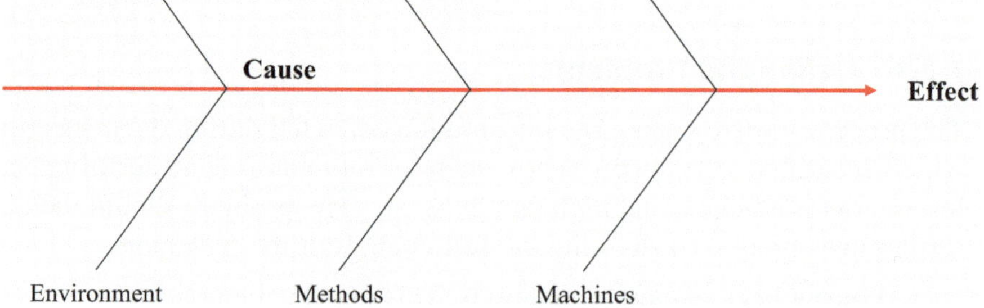

Figure 7.1 Cause and effect chart.

corrective actions. Root-cause analysis and the determination of deficiency and energy waste type, cause, a lead to accurate, efficient, and cost-effective corrective actions.

A sample equipment list, with defined deficiencies (i.e., motor consumption), are defined in Chapter 9. The table lists down all possible causes against a certain operational problem.

7.3.3 How to Use Fishbone Diagram

The fishbone diagram swims to the right. The effect is on the right and can be either a problem or opportunity. For ECM, it is a problem. The top or bottom with personnel, machines, materials, methods, environment, and measurements are called the major bones of the fishbone. The major bones' titles are not usually specified. They can come from a process or these lists:

- **Six M's** – Machines, methods, money, material, measurement, and management.
- **Five P's** – people, procedure, program, process, and personnel.
- **One S** – System.
- **One E** - Environment.
- **Three T's** – Tools, techniques, and training.

The major bones shown in Figure 7.2 are common in determining root causes for problems experienced in ECM. Normally, four major bones are sufficient to add sufficient bones to the major bones to identify the potential root causes. Visio has a fishbone diagram and is easy to use. Also, draw the skeleton (effect and the four major bones) on a white board and then the team or one supervisor or technician can write a possible root cause on a yellow 3M Company's "post it." And then stick it on the appropriate major bone and keep doing this until no other possible root causes can be thought of by the participant(s). Next, each possible root cause is evaluated, and the ones that are the strongest possibilities are circled. They are then validated using data that were available or a root-cause matrix.

Often if insufficient people are available to accomplish all the maintenance and maintenance management duties, someone may be chasing too may rabbits. Running this way and then that way and not getting anything done. To discover potential root causes, they decided to use a root-cause analysis, specifically a fishbone diagram to identify why they are having this problem. Figure 7.2 shows this.

Notice all the small bones and how the arrows go into the bones and then the major bones arrows point and go into the effect. The causes flow

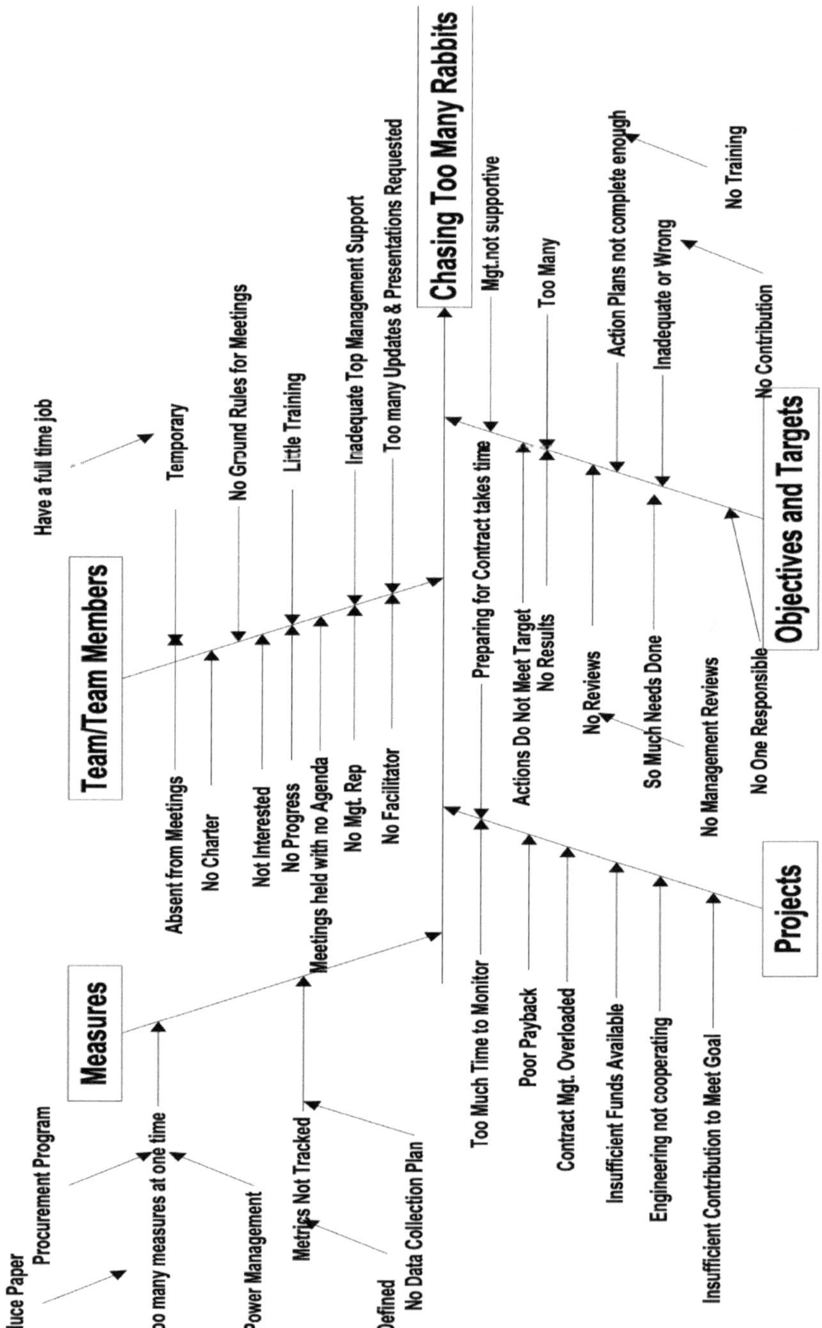

Figure 7.2 Fishbone diagram.

in the effect. This is why it is called a cause and effect diagram. There are normally many small bones and only four major bones. In this example, the major bones selected were team/team members, objectives and targets, projects, and measures. Each major bone has several small bones.

The energy wastes that are captured during the implementation of the ECM model can result. Later in this book, an example of energy wastes might have multiple causes. Chapter 9 lists a detailed list of each problem, cause, and effect and suggests ways to eliminate those causes.

Energy waste causes during equipment operation can be one of the following types (note that this list is not comprehensive):

1. Running when it should be turned off.
2. Having to work harder.
3. Wear.
4. Corrosion.
5. Leaks.
6. Stuck in the open or closed position.
7. Not performing.
8. Overperforming.
9. Overheating.
10. Waste build up.
11. Incorrect settings.
12. Switch or sensor failure (broke or seized).

In the future, as organizations implement ECM, this list should grow. In applying ECM later, examples will be shown for each of these energy wastes or excessive uses of energy. ECM includes routine inspections and then fixing any problems encountered.

8

ECM Process – Identifying Corrective/ Preventive Action and Cost Effectiveness

8.1 Step 5: Identifying Corrective/Preventive Action and Cost Effectiveness

When the root cause of a certain equipment performance deficiency is known, the right corrective action and repair can be implemented. Energy centered maintenance model focuses on identifying any operational deficiencies or energy waste; therefore, the corrective action may include specific repairs or replacements. Therefore the cost effectiveness of the repair needs to be determined prior to conducting the corrective action.

8.2 Identifying Corrective/Preventive Action

Improving equipment efficiency requires proactive involvement in finding the root cause of a deficiency or energy waste and then instituting corrective action for improvements. Corrective/preventive action is an action taken to repair, improve, or restore an equipment failure or performance and to prevent it from reoccurrence. The action is decided based on the findings of the root-cause analysis of the problem and the associated cause and effect results which aim to prevent the problem from appearing again.

When the maintenance personnel conducts ECM inspections and they find that certain equipment is underperforming or wasting energy, an immediate corrective action may be undertaken, or it may be scheduled. If the deficiency is critical for the facility operation or equipment's energy waste is high, corrective action should be planned and schedule within 0–48 hrs from the time of reporting the problem.

The corrective actions may then be amended in the normal maintenance program by adding new maintenance tasks that provide a proactive approach to prevent the problem from reoccurring; in this case, the corrective action will be conducted in a proactive manner which makes it a preventive action. Adjustments can be made to the balanced proactive maintenance strategy approach to best suit the equipment.

Energy centered maintenance is a balanced integration of preventive, predictive, and proactive maintenance strategies that emphasize the elimination of energy wastes, thereby controlling energy consumption and equipment efficiency. The integration between those maintenance strategies with root-cause analysis and corrective/preventive actions leads to improving the maintenance regime and eliminating the problem of reoccurrence.

An example of a corrective action in ECM model is a fan that is delivering lower airflow rate than originally commissioned; the cause maybe a worn bearing and root cause is the low level of lubrication. In this example, the corrective action is to replace the bearing, while the preventive action is to adjust the normal maintenance program to inspect the bearings and the level of lubrication as part of the planned maintenance.

Identifying corrective action will help in improving the overall balanced, proactive maintenance strategy in the following:

- Identifying what needs to be done to restore equipment performance and what the expected results are.
- Identifying preventive actions to prevent the problem from being happening again.
- Identifying which maintenance process needs improvement.
- Identifying if new processes need to be implemented.
- Identifying new training for maintenance personnel.
- Identifying all costs associated with implementation and determining cost effectiveness.

8.3 Identifying Cost Effectiveness

Maintenance engineers and energy managers are responsible for verifying that appropriate and effective corrective actions have been taken in a cost-effective manner.

Before implementing the corrective and preventive actions, it is required to consider the cost effectiveness of the maintenance actions. Some actions may require no or low cost for implementation (i.e., replacing temperature sensor), while other actions may require an investment cost (i.e., replacing motor).

The cost effectiveness of the corrective and preventive actions is calculated based on the cost of implementation compared to the energy cost saved. If the energy cost saved is higher than the corrective and preventive actions cost, the energy centered maintenance implementation is cost-effective.

Cost effectiveness should be calculated for each ECM tasks as well as corrective and preventive actions; calculating cost effectiveness should count for all maintenance activities costs associated with those actions, as well as potential energy reduction as defined in a certain period.

The maintenance costs include multiple elements such as:

- Time of maintenance.
- Labor cost.
- Material and consumable cost.
- Equipment cost.
- Calibration cost.
- Spare parts cost.

Energy conservation is a basic unit of ECM. It is cost-effective in that it does not include high labor costs and materials. Most conservation actions are relatively inexpensive or would already be purchased if an energy conservation program were implemented.

Energy saved could be calculated by different ways for each type of equipment by itself. For example, energy saved by enhancing motor efficiency equals the difference in motor's kWh before and after enhancement. This difference then can be converted to cost saving in a defined period (can be months, years, or the life cycle of the equipment) and can be compared to maintenance cost.

The main aim of energy centered maintenance model is to assist the organization in its overall energy program to reduce its energy use and to ensure efficient cost-effective operation. If the cost of ECM maintenance actions is less than the impact of the energy waste of the equipment, the appropriate corrective and preventive maintenance activities should be implemented.

The following formula should assist the maintenance personnel to identify the cost effectiveness of the corrective/preventative actions that need to be implemented to restore the equipment efficiency (i.e., motor power consumption).

$$Cost\ Effectivness(\%) = \frac{Cost\ of\ Periodic\ Energy\ Saved - Cost\ of\ Maintenance\ Tasks}{Cost\ of\ Periodic\ Energy\ Saved} \times 100\%$$

If the cost-effectiveness value is equal or more than zero, then it is cost-effective to conduct the energy centered maintenance tasks and the associated corrective and preventive actions.

On the other hand, if the value is in minus (less than zero), then the cost of maintenance is higher than the cost of energy saved. In this case, the maintenance personnel should account for other factors in determining whether to conduct the ECM tasks or not.

Factors like the contribution of this equipment energy saving to the total building energy saving, impact of the low-performance cost of business operation, and the impact of low-performance cost on occupant's comfort level will be required to evaluate the effectiveness and the need of returning the equipment to its optimum operational efficiency.

8.4 Restoring Equipment Efficiency

Restoring equipment efficiency aims to operate the equipment in a satisfactory manner. Satisfactory performance implies that specific criteria must be established to describe what is considered as satisfactory operation. Satisfactory operation is achieved when the equipment is operating as per its design intent and as original commissioned or in a more efficient manner.

Corrective actions are required to restore the performance of a non-performing equipment; preventive actions prevents the problem to happen again. It is necessary to verify that the corrective action was effective, not only in eliminating the cause of the deficiency but also in restoring the equipment's performance. Whatever corrective action is taken, ensure that the original problem is fixed and that no new deficiency is introduced.

Energy centered maintenance focuses on an environment of actions in which equipment efficiency is the main concern. Once ECM inspection is done, and the problem is defined, corrective actions should be implemented to restore the performance of the equipment. Restoring equipment performance is focusing on the following outputs:

- Maximize equipment's operational efficiency.
- Restoring equipment's energy efficiency.
- Reducing energy waste by the equipment.
- Lowers operation cost by reducing energy consumption.
- Restoring the original performance of the equipment.
- Improves equipment quality.
- Ensuring that equipment is delivering its intended performance.
- Understanding the effect of equipment age, operating environment on its performance.

- Obtaining data for continuous improvement and high operational effectiveness.
- Corrective/preventive actions are required to restore the performance of a failing equipment. It is necessary to verify that the corrective action was effective, not only in eliminating the cause of the failure but also in restoring the equipment's performance. Whatever correction/preventive actions are taken, ensure that the original problem is fixed and that no new problems are introduced.

The root-cause analysis process and identification of corrective actions decide what actions are needed to restore the efficiency of the equipment and its performance. Cost effectiveness decides which measures are most cost-effective to be implemented. Restoring equipment efficiency and performance may require a full design review that defines how the performance will be restored. For example, replacing fan motor or rewinding it.

Implementation of energy centered maintenance tasks in a proactive manner allows the equipment to perform in an efficient manner for a given time when used under specified operating conditions in a given environment. Implementation of ECM tasks results in restoring equipment performance to its original condition so that the equipment can continue to be used for its intended function.

9

ECM Process – Updating Preventative Maintenance Plans

9.1 Step 6: Updating PM Plans on CMMS

Once the energy centered maintenance approach has been established, regular use of the maintenance program should be implemented in terms of planned preventive maintenance as well as predictive maintenance. The new program should be updated on the balanced proactive maintenance strategy to establish a proactive system that reviews the energy-efficient data of the equipment which helps to identify system operational performance before a significant deficiency occurs.

The new maintenance program should be updated in the maintenance management system (CMMS) to ensure that all energy centered maintenance tasks that are defined during ECM inspection are now part of the preventive maintenance plans and predictive maintenance practice.

9.2 What is CMMS?

CMMS is sometimes called CMMIS with the "I" being "Information" which describes what it is. It is a computerized maintenance management system that contains an organization's maintenance activities. The "IS" stands for the information system. It is also referred to as the organization's facility maintenance program.

A CMMS software package maintains a computer database of information about an organization's maintenance program and operations. This information is intended to help improve maintenance workers' efficiency and productivity. For example, determining which equipment requires maintenance – what, when, and which supply rooms contain the parts or materials needed for a particular work order. And to help the management make better decisions (for example, comparing the cost of equipment failure versus preventive maintenance for each machine), CMMS data may also be used to show regulatory compliance and compliance with an ISO requirement.

The following identifies what the CMMS should be able to do:

- Address all resources involved in maintenance.
- Maintain maintenance inventory and storeroom location.
- Record and maintain work history of all types of maintenance PM, predictive, emergency, and corrective.
- Include work tasks and frequencies for each craft.
- Effectively interface and communicate with related and supporting systems ranging from work generation through work performance, evaluation, and performance reporting.
- Provide feedback information for analysis and decision-making.
- Reduce costs through effective maintenance planning and execution.

A modern CMMS meets all these requirements and assists the facilities maintenance manager with, planning, scheduling, control, performance, evaluation, and reporting. CMMS will also maintain historical data for management use and provide meaningful maintenance metrics. Therefore, CMMS provides a work order system, asset management, inventory/purchasing, PM management, work scheduling, and management reports. CMMS is common in the manufacturing industry, both government and civilian facilities, fleet, service providers, oil and gas, and other industries. CMMS software packages can be either web-based (hosted by the company selling the product on an outside server) or LAN-based (the organization buying the software hosts the product on their server).

CMMS improves the mechanic's or a technician's wrench time, enhances spare parts inventory, and streamlines procurement of parts and materials. CMMS benefits are:

- Easy to use.
- Quick to implement maintenance program.
- Minimizes downtime and improves productivity.
- Provides maintenance records and history that serves as compliance proof.

CMMS provides preventative maintenance scheduling, work orders, work or service requests, inventory control, predictive maintenance, maintenance reports, and other functions. The CMMS user interface allows for a quick setup and easy data conversion, and the vendor provides training for the users.

CMMS vendors claim that CMMS reduces costs and asset downtime and increases productivity in less than a month. They claim that it will extend the life cycle for your facility, decrease your liability, and lower your operating costs without any significant upfront investment.

It provides online planned maintenance scheduling that helps generate, schedule, and manage recurring tasks which are the heart of ECM. Some systems allow sending work order information to maintenance crews in the field, enabling them to receive and complete tasks away from their shop. CMMS is excellent at scheduling jobs, assigning personnel, earmarking materials, recording maintenance costs, and tracking information such as the cause of the problem (if any), downtime that occurred (if any), and suggestions for future action. The CMMS schedules preventive maintenance based on maintenance plans. Different CMMS software packages use different techniques for highlighting when a PM job or task should be performed.

CMMS keeps track of preventive maintenance jobs, including step-by-step instructions. This action is critical for ECM to be successful. CMMS provides an online work order management system that streamlines your work order process, including work request generation, progress and completion status tracking and reports, and reporting of essential maintenance management data and information.

9.3 Updating PM Plans on CMMS

Updating CMMS and maintenance program with specific energy centered maintenance tasks provide vital information to the maintenance and operations personnel that enables them to maintain efficiency and mitigate performance degradation as well as minimize or eliminate energy wastes associated with the equipment operation.

Energy centered maintenance plans should be created for each energy critical equipment, and it should be integrated with regular reliability maintenance plans. The PM plans should be effective, efficient, and safe to perform based on the strategy (predictive and/or preventive) applied to the equipment. PM plans are also defined as job plans; job plans must be written for every equipment based on energy criticality.

Updating PM plans requires the following:

- Identifying ECM tasks and frequency.
- Identifying and matching the appropriate skill sets to the tasks.
- Identifying the appropriate materials to the tasks.
- Identifying the appropriate tools/special equipment to the tasks.
- Identifying all other resources needed to perform the job.
- Uploading completed job plans to current CMMS.

The job plans provide all the details regarding safety, environmental, and regulatory issues, as well as the operations, required downtime, affected components/systems, materials, labor, and tools required to do the work. The procedural part of the plan contains a task or a logical sequence of tasks, while each task consists of some related steps.

The fundamental objective is to conduct efficient, safe, and standardized energy centered maintenance job plans as part of regular reliability maintenance plan, which ensures proper quality PM procedures that maintain the ongoing efficiency of equipment.

Integrating ECM job plans with reliability job plans will enhance the productivity of the service provider who is conducting the tasks on-site. Job plans should also be scheduled and tracked to identify the effectiveness of the service provider. Therefore, job plans should be updated in the company's CMMS system.

The use of CMMS will help to develop a site-specific energy centered maintenance program based on equipment functionality, designed to enhance equipment efficiency, make the best use of maintenance resources, and provide correct data for additions or revisions to the existing or new maintenance programs.

9.4 Planning and Scheduling Next Inspection

When energy centered maintenance job plans are developed, the FM team shall plan and schedule the next inspection to ensure that a proactive approach is taken to maintenance equipment efficiency.

Maintenance planning and scheduling is a significant improvement strategy for maintenance operations and is the single most effective procedure to increase craft labor productivity, effectiveness, and quality. Planning and scheduling work in advance provides the facility and the ability to control maintenance activities, reduce costs, reduce the risk of equipment performance deficiency, and improve productivity.

Planning and scheduling emphasize the need to plan and schedule energy centered maintenance activities, to maximize the wrench time of the craftspeople doing the work, and to create a situation whereby the people who do the tasks arrive at a specified work site fully prepared, with all the necessary instructions, permits, clearances, materials, tools, and equipment to undertake the work order tasks associated with a particular job.

This is a significant improvement strategy for maintenance operations because the work is planned, and the right resources are coordinated and are available to do the job correctly the first time. The difference between planning and scheduling is:

- Effective planning takes into consideration all of the factors involved in doing the job, along with the sequence in which the factors come into play. By coordinating available resources, effective planning facilitates establishing minimum time and optimum cost work methods.
- Efficient scheduling assures a balanced flow of work to the shops by maintaining a proper balance between work capacity and workload.

When the join plans are planned and scheduled, the ECM inspections will take place on a regular basis as part of reliability maintenance plans. Executing maintenance job plans will maintain the efficiency of the equipment and will minimize or eliminate energy waste.

Each site should analyze the potential benefits to be derived from planning and scheduling energy centered maintenance inspection and should develop the plan to implement a planning and scheduling function. This will ensure that:

- ECM inspection is identified and prioritized.
- ECM inspection is accurately planned.
- ECM inspection is scheduled at the right time.
- ECM inspection is assigned and executed in a timely manner.

Energy centered maintenance inspections will help the FM personnel to identify any potential improvements that can be conducted to a certain equipment to ensure that it is functioning as originality designed, with high efficiency, and no energy waste.

9.5 Sample Problem, Cause, Effect, and Corrective/ Preventive Actions

In relation to previous sections, this section provides sample cases for the main energy consuming equipment in buildings, with the possible causes of low equipment's eprfomance and the required correct/preventative actions. Information is provided in tables 9.1 to 9.34 for each type of equipment listed in tables 6.3-6.36 previously.

9.5.1 Heating, Ventilation, and Air Conditioning System

Table 9.1 Air handling unit.

Equipment type: air handling units

Preventive maintenance

Maintenance inspection	Acceptable performance	Problem/effect in case of low performance	Some possible causes in case of low performance – *SAMPLE*	Corrective/preventive action
Measure airflow rate (m^3/hr)	Current value = ±5% of testing and commissioning value	• Problem: low airflow • Effect: low cooling in served area	• Increase in static pressure • Worn bearing • Worn belt	• Check whether any dampers are closed, filters clogged, etc. • Replace bearing
Check motor's full load current (Amps)	Match value on data plate		• Low motor efficiency	• Replace belt • Check winding and restore
Measure cooling coil performance on full load (air off-coil temperature (°C)	Current value = ±5% of testing and commissioning value	• Problem: high off-coil temperature • Effect: low cooling in served area	• Dirty cooling coil • Blocked strainer • High supply chilled water temperature	• Clean coil • Clean strainer • Check and reduce chilled temperature
Measure cooling coil performance (Pressure drop, PSI)	Current value = ±5% of cooling coil selection pressure drop	• Problem: high-pressure drop • Effect: higher electrical consumption in pump's motor	• Scales accumulated inside the coil • Increased chilled water flow rate inside the coil than T&C	• Clean or replace coil • Check DRV and PICV setting and check DPT index value
Measure variable frequency drive effectiveness	In-voltage is the same value as that of main source voltage, out-voltage = %frequency	• Problem: higher/lower required power • Effect: higher energy consumption. Less efficiency	• Drive defected • Meters not calibrated • Ambient temperature • Load defected	• Check drive cooling fans and fix it • Maintain the ambient temperature to accepted value • Check meters calibration and recalibrate • Check load parameter and adjust
Measure two-way/three-way control valves response to space temperature	100% functionality with design intent	• Problem: stuck open valve • Effect: overcooling, higher consumption	• Stuck valve – open or close • Loosed signal from BMS	• Restore valve to automatic • Restore signal and logic

Measure chilled water temperature difference (delta T) °C	100% ± 5% achieving desired delta T	• Problem: low delta T • Effect: higher consumption	• Blocked coils • Stuck opened valve • Low set points (temperature or system DPT)	• Clean coils • Restore valve to auto mode • Adjust set points

Predictive maintenance

Maintenance task	Acceptable performance	Problem/effect in case of low performance	Some possible causes in case of low performance – *SAMPLE*	Corrective/preventive action
Measure airflow rate (m^3/hr)	Operating value within predefined alarm limits	• Problem: low airflow • Effect: low cooling in served area	• Increase in static Pressure • Worn bearing • Worn belt • Low motor efficiency	• Check whether any dampers are closed, filters clogged, etc. • Replace bearing • Replace belt • Check winding and restore
Measure air off-coil temperature (°C)	Operating value within predefined alarm limits	• Problem: high off-coil temperature • Effect: low cooling in served area	• Dirty cooling coil • Blocked strainer • High supply chilled water temperature	• Clean coil • Clean strainer • Check and reduce chilled temperature
Measure fan's motor power consumption (kWh)	Operating value within predefined alarm limits	• Problem: high power consumption • Effect: low cooling in served area	• Increase in static pressure • Low motor efficiency • Damaged winding	• Check whether any dampers are closed, filters clogged, etc. • Replace bearing • Replace belt • Check winding and restore
Measure chilled water temperature difference (delta T) °C	Operating value within predefined alarm limits	• Problem: low delta T • Effect: higher consumption	• Blocked coils • Stuck opened valve • Low set points (temperature or system DPT)	• Clean coils • Restore valve to auto mode • Adjust set points

Table 9.2 Fan coil units.

Equipment type: fan coil units

Preventive maintenance

Maintenance inspection	Acceptable performance	Problem/effect in case of low performance	Some possible causes in case of low performance – *SAMPLE*	Corrective/preventive action
Measure airflow rate (m^3/hr)	Current value = ±5% of testing and commissioning value	• Problem: low airflow • Effect: low cooling in served area	• Increase in static pressure • Worn bearing • Worn belt • Low motor efficiency	• Check whether any dampers are closed, filters clogged, etc. • Replace bearing • Replace belt • Check winding and restore
Check motor's full load current (Amps)	Match value on data plate			
Measure cooling coil performance on full load (air off-coil temperature °C)	Current value = ±5% of testing and commissioning value	• Problem: high off-coil temperature • Effect: low cooling in served area	• Dirty cooling coil • Blocked strainer • High supply chilled water temperature	• Clean coil • Clean strainer • Check and reduce chilled temperature
Measure cooling coil performance (pressure drop, PSI)	Current value = ±5% of cooling coil selection pressure drop	• Problem: high-pressure drop • Effect: higher electrical consumption in pump's motor	• Scales accumulated inside the coil • Increased chilled water flow rate inside the coil than T&C	• Clean or replace coil • Check DRV & PICV setting and check DPT index value
Measure variable frequency drive effectiveness	In-voltage is same value of main source voltage, out-voltage = % frequency	• Problem: higher/lower required power • Effect: higher energy consumption. Less efficiency	• Drive defected • Meters not calibrated • Ambient temperature • Load defected	• Check drive cooling fans and fix it • Maintain the ambient temperature to accepted value • Check meters calibration and recalibrate • Check load parameter and adjust
Measure two-way/three-way control valves response to space temperature	100% functionality with design intent	• Problem: stuck open valve • Effect: overcooling, higher consumption	• Stuck valve – open or close • Loosed signal from BMS	• Restore valve to automatic • Restore signal and logic

Measure chilled water temperature difference (delta T) °C	100% ± 5% achieving desired delta T	• Problem: low delta T • Effect: higher consumption	• Blocked coils • Stuck opened valve • Low set points (temperature or system DPT)	• Clean coils • Restore valve to auto mode • Adjust set points

Predictive maintenance

Maintenance task	Acceptable performance	Problem/effect in case of low performance	Some possible causes in case of low performance – *SAMPLE*	Corrective/preventive action
Measure airflow rate (m^3/hr)	Operating value within predefined alarm limits	• Problem: low airflow • Effect: low cooling in served area	• Increase in static pressure • Worn bearing • Worn belt • Low motor efficiency	• Check whether any dampers are closed, filters logged, etc. • Replace bearing • Replace belt • Check winding and restore
Measure air off-coil temperature (°C)	Operating value within predefined alarm limits	• Problem: high off-coil temperature • Effect: low cooling in served area	• Dirty cooling coil • Blocked strainer • High supply chilled water temperature	• Clean coil • Clean strainer • Check and reduce chilled temperature
Measure fan's motor power consumption (kWh)	Operating value within predefined alarm limits	• Problem: high power consumption • Effect: low cooling in served area	• Increase in static pressure • Low motor efficiency • Damaged winding	• Check whether any dampers are closed, filters clogged, etc. • Replace bearing • Replace belt • Check winding and restore
Measure chilled water temperature difference (delta T) °C	Operating value within predefined alarm limits	• Problem: low delta T • Effect: higher consumption	• Blocked coils • Stuck opened valve • Low set points (temperature or system DPT)	• Clean coils • Restore valve to auto mode • Adjust set points

Note: Predictive maintenance measures for fan coil units may be implemented for high energy consumption FCUs. It may not be feasible to implement it on all FCUs within the facility.

Table 9.3 Energy recovery units (i.e., heat wheels).

Equipment type: energy recovery unit

Preventive maintenance

Maintenance inspection	Acceptable performance	Problem/effect in case of low performance	Some possible causes in case of low performance – *SAMPLE*	Corrective/preventive action
Measure airflow rate (m^3/hr)	Current value = ±5% of testing and commissioning value	• Problem: low airflow • Effect: low cooling in served area	• Increase in static pressure • Worn bearing • Worn belt • Low motor efficiency	• Check whether any dampers are closed, filters clogged, etc. • Replace bearing • Replace belt • Check winding and restore
Check motor's full load current (Amps)	Match value on data plate			
Measure energy recovery performance on full load (air off energy recovery unit temperature °C)	Current value = ±5% of testing and commissioning value	• Problem: high supply temperature • Effect: insufficient thermal comfort	• Clogged coil • Damaged coil • Dirty coil	• Clean coil • Repair coil
Measure cooling coil performance (pressure drop, PSI)	Current value = ±5% of cooling coil selection pressure drop	• Problem: high-pressure drop • Effect: higher electrical consumption in pump's motor	• Scales accumulated inside the coil • Increased chilled water flow rate inside the coil than T&C	• Clean coil • Maintain flow rate as per design

Predictive maintenance

Maintenance task	Acceptable performance	Problem/effect in case of low performance	Some possible causes in case of low performance – *SAMPLE*	Corrective/preventive action
Measure airflow rate (m^3/hr)	Operating value within predefined alarm limits	• Problem: low airflow • Effect: low cooling in served area	• Increase in static pressure • Worn bearing • Worn belt • Low motor efficiency	• Check whether any dampers are closed, filters clogged, etc. • Replace bearing • Replace belt • Check winding and restore
Measure air off energy recovery unit temperature (°C)	Operating value within predefined alarm limits	• Problem: high supply temperature • Effect: insufficient thermal comfort	• Clogged coil • Damaged coil • Dirty coil	• Clean coil • Repair coil
Measure fan's motor power consumption (kWh)	Operating value within predefined alarm limits	• Problem: high power consumption • Effect: low cooling in served area	• Increase in static pressure • Low motor efficiency • Damaged winding	• Check whether any dampers are closed, filters clogged, etc. • Replace bearing • Replace belt • Check winding and restore

Table 9.4 Boilers.

Equipment type: boilers

Preventive maintenance

Maintenance inspection	Acceptable performance	Problem/effect in case of low performance	Some possible causes in case of low performance – *SAMPLE*	Corrective/preventive action
Measure fuel combustion efficiency	±5% of original efficiency	• Problem: low efficiency • Effect: insufficient supply water temperature/high energy consumption	• Air fuel ratio is not balanced • Combustion process is not efficient • Supply of high amount of water, more than design flow rate	• Test and calibrate air/fuel ratio • Do all required cleaning at burners, heat exchangers, etc. • Supply water flow rate as per design and check on a regular basis
Inspect steam leakage	No steam leakage	• Problem: steam leakage • Effect: high energy consumption	• Leakage in tubes, plates, etc.	• Identify leakage location and restore • Check pressure on a regular basis

Predictive maintenance

Maintenance task	Acceptable performance	Problem/effect in case of low performance	Some possible causes in case of low performance – *SAMPLE*	Corrective/preventive action
Outlet water temperature	Operating value within predefined alarm limits	• Problem: low water temperature • Effect: insufficient supply water temperature/high energy consumption	• Air fuel ratio is not balanced • Combustion process is not efficient • Supply of high amount of water, more than design flow rate	• Test and calibrate air/fuel ratio • Do all required cleaning at burners, heat exchangers, etc. • Supply water flow rate as per design and check on regular basis
Primary system water pressure	Operating value within predefined alarm limits	• Problem: steam leakage • Effect: high energy consumption	• Leakage in tubes, plates, etc.	• Identify leakage location and restore • Check pressure on regular basis

Table 9.5 Pumps.

Equipment type: pumps

Preventive maintenance

Maintenance inspection	Acceptable performance	Problem/effect in case of low performance	Some possible causes in case of low performance – *SAMPLE*	Corrective/preventive action
Measure full speed water flow rate (gpm, lps)	Match value on data plate	• Problem: low water flow rate • Effect: insufficient water delivery – based on function.	• Clogged strainer • Air clogged in the system • Cavitation • Low motor speed • Torn impeller	• Clean strainer • Evacuate air from system • Fix motor winding
Check motor's full load current (Amps)	Match value on data plate	• Problem: low water flow • Effect: pump not capable of functioning as designed, high energy consumption	• Increase in the pump head • Worn bearing • Blocked strainers • Low motor efficiency	• Verify pump head and check impeller • Replace bearing • Clean strainer • Check winding and restore
Measure variable frequency drive effectiveness	In-voltage is the same value of main source voltage, out-voltage = % frequency	• Problem: higher/lower required power • Effect: higher energy consumption. Less efficiency	• Drive defected • Meters not calibrated • Ambient temperature • Load defected	• Check drive cooling fans and fix it • Maintain the ambient temperature to accepted value • Check meters calibration and recalibrate • Check load parameter and adjust

Predictive maintenance

Maintenance task	Acceptable performance	Problem/effect in case of low performance	Some possible causes in case of low performance – *SAMPLE*	Corrective/preventive action
Measure water flow rate (gpm, lps)	Operating value within predefined alarm limits	• Problem: low water flow rate • Effect: insufficient water delivery – based on function	• Clogged strainer • Air clogged in the system • Cavitation • Low motor speed • Torn impeller	• Clean strainer • Evacuate air from system • Fix motor winding
Measure pump's motor power consumption (kWh)	Operating value within predefined alarm limits	• Problem: kWh more than normal range • Effect: high energy consumption	• Damaged bearing • Blocked strainer • Damaged PF	• Replace bearing • Clean strainer • Replace PF
Motor running current	Operating value within predefined alarm limits	• Problem: kWh more than normal range • Effect: high energy consumption	• Damaged bearing • Blocked strainer • Damaged PF	• Replace bearing • Clean strainer • Replace PF

Table 9.6. Close control units.

Equipment type: close control units				
Preventive maintenance				
Maintenance inspection	**Acceptable performance**	**Problem/effect in case of low performance**	**Some possible causes in case of low performance – *SAMPLE***	**Corrective/preventive action**
Measure airflow rate (m^3/hr)	Current value = ±5% of testing and commissioning value	• Problem: low airflow • Effect: low cooling in served area	• Increase in static pressure • Worn bearing • Worn belt	• Check whether any dampers are closed, filters clogged, etc. • Replace bearing
Check fan motor's full load current (Amps)	Match value on data plate		• Low motor efficiency	• Replace belt • Check winding and restore
Check DX unit compressor full load current (Amps)	Match value on data plate	• Problem: high Amps • Effect: high energy consumption	• Low refrigerant level • Low motor efficiency	• Check and maintain refrigerant level • Check winding and restore
Measure cooling coil performance on full load (air off-coil temperature °C)	Current value = ±5% of testing and commissioning value	• Problem: high off-coil temperature • Effect: low cooling in served area	• Dirty cooling coil • Blocked strainer • High supply chilled water temperature	• Clean coil • Clean strainer • Check and reduce chilled temperature
Measure cooling coil performance (pressure drop, PSI)	Current value = ±5% of cooling coil selection pressure drop	• Problem: high pressure drop • Effect: higher electrical consumption in pump's motor	• Scales accumulated inside the coil • Increased chilled water flow rate inside the coil than T&C	• Clean or replace coil • Check DRV and PICV setting and check DPT index value
Measure variable frequency drive effectiveness	In-voltage is the same value as that of main source voltage, out-voltage = % frequency	• Problem: higher/lower required power • Effect: higher energy consumption. Less efficiency	• Drive defected • Meters not calibrated • Ambient temperature • Load defected	• Check drive cooling fans and fix it • Maintain the ambient temperature to accepted value • Check meters calibration and recalibrate • Check load parameter and adjust

Table 9.6 Continued

Equipment type: close control units				
Preventive maintenance				
Maintenance inspection	**Acceptable performance**	**Problem/effect in case of low performance**	**Some possible causes in case of low performance – *SAMPLE***	**Corrective/preventive action**
Measure two-way/three-way control valves response to space temperature	100% functionality with design intent	• Problem: stuck open valve • Effect: overcooling, higher consumption	• Stuck valve – open or close • Loosed signal from BMS	• Restore valve to automatic • Restore signal and logic
Measure chilled water temperature difference (delta T) °C	100% ± 5% achieving desired delta T	• Problem: low delta T • Effect: higher consumption	• Blocked coils • Stuck opened valve • Low set points (temperature or system DPT)	• Clean coils • Restore valve to auto mode • Adjust set points
Measure DX unit energy efficiency ratio (EER)	±5% value on data plate	• Problem: lower EER • Effect: higher energy consumption	• Low refrigerant level • Low motor efficiency	• Check and maintain refrigerant level • Check winding and restore

Predictive maintenance				
Maintenance task	**Acceptable performance**	**Problem/effect in case of low performance**	**Some possible causes in case of low performance – *SAMPLE***	**Corrective/preventive action**
Measure airflow rate (m^3/hr)	Operating value within predefined alarm limits	• Problem: low airflow • Effect: low cooling in served area	• Increase in static pressure • Worn bearing • Worn belt • Low motor efficiency	• Check whether any dampers are closed, filters clogged, etc. • Replace bearing • Replace belt • Check winding and restore
Measure air off-coil temperature (°C)	Operating value within predefined alarm limits	• Problem: high off-coil temperature • Effect: low cooling in served area	• Dirty cooling coil • Blocked strainer • High supply chilled water temperature	• Clean coil • Clean strainer • Check and reduce chilled temperature
Check DX unit compressor full load current (Amps)	Operating value within predefined alarm limits	• Problem: high Amps • Effect: high energy consumption	• Low refrigerant level • Low motor efficiency	• Check and maintain refrigerant level • Check winding and restore
Measure chilled water temperature difference (delta T) °C	Operating value within predefined alarm limits	• Problem: low delta T • Effect: higher consumption	• Blocked coils • Stuck opened valve • Low set points (temperature or system DPT)	• Clean coils • Restore valve to auto mode • Adjust set points

Table 9.7 Fans.

Equipment type: fans

Preventive maintenance

Maintenance inspection	Acceptable performance	Problem/effect in case of low performance	Some possible causes in case of low performance – *SAMPLE*	Corrective/preventive action
Measure airflow rate (m^3/hr)	Match value on data plate	• Problem: low airflow • Effect: low cooling in served area	• Increase in static pressure • Worn bearing • Worn belt • Low motor efficiency	• Check whether any dampers are closed, filters clogged, etc. • Replace bearing • Replace belt • Check winding and restore
Check motor's full load current (Amps)	Match value on data plate			
Measure variable frequency drive effectiveness	In-voltage is the same value as that of main source voltage, out-voltage = % frequency	• Problem: higher/lower required power • Effect: higher energy consumption. Less efficiency	• Drive defected • Meters not calibrated • Ambient temperature • Load defected	• Check drive cooling fans and fix it • Maintain the ambient temperature to accepted value • Check meters calibration and recalibrate • Check load parameter and adjust

Predictive maintenance

Maintenance task	Acceptable performance	Problem/effect in case of low performance	Some possible causes in case of low performance – *SAMPLE*	Corrective/preventive action
Measure airflow rate (m^3/hr)	Operating value within predefined alarm limits	• Problem: low airflow • Effect: low cooling in served area	• Increase in static pressure • Worn bearing • Worn belt • Low motor efficiency	• Check whether any dampers are closed, filters clogged, etc. • Replace bearing • Replace belt • Check winding and restore
Measure pump's motor power consumption (kWh)	Operating value within predefined alarm limits	• Problem: high power consumption • Effect: low cooling in served area	• Increase in static pressure • Low motor efficiency • Damaged winding	• Check whether any dampers are closed, filters clogged, etc. • Replace bearing • Replace belt • Check winding and restore

Table 9.8 Cooling towers.

Equipment type: cooling towers

Preventive maintenance

Maintenance inspection	Acceptable performance	Problem/effect in case of low performance	Some possible causes in case of low performance – *SAMPLE*	Corrective/preventive action
Check fan motor's full load current (Amps)	Match value on data plate	• Problem: high Amps • Effect: high consumption, low airflow rate extraction	• Blocked louvers • Worn bearing • Damaged winding	• Clean louvers • Replace bearing • Fix winding
Cooling tower range (in–out water temperature) – full load (°C)	Match value on data plate	• Problem: low water range • Effect: high outlet temperature	• Blocked louvers • Higher amount of water discharged from sprinklers • Damaged fan blades	• Clean and fix louvers • Check and calibrate water from sprinklers • Check and replace fan blades
Measure variable frequency drive effectiveness	In-voltage is the same value as that of main source voltage, out-voltage = % frequency	• Problem: higher/lower required power • Effect: higher energy consumption. Less efficiency	• Drive defected • Meters not calibrated • Ambient temperature • Load defected	• Check drive cooling fans and fix it • Maintain the ambient temperature to accepted value • Check meters calibration and recalibrate • Check load parameter and adjust

Predictive maintenance

Maintenance task	Acceptable performance	Problem/effect in case of low performance	Some possible causes in case of low performance – *SAMPLE*	Corrective/preventive action
Check fan motor's current (Amps)	Operating value within predefined alarm limits	• Problem: high Amps • Effect: high consumption, low airflow rate extraction	• Blocked louvers • Worn bearing • Damaged winding	• Clean louvers • Replace bearing • Fix winding
Cooling tower range (in–out water temperature) – full load (°C)	Operating value within predefined alarm limits	• Problem: low water range • Effect: high outlet temperature	• Blocked louvers • Higher amount of water discharged from sprinklers • Damaged fan blades	• Clean and fix louvers • Check and calibrate water from sprinklers • Check and replace fan blades

Table 9.9 Air cooled chillers.

Equipment type: air cooled chillers

Preventive maintenance

Maintenance inspection	Acceptable performance	Problem/effect in case of low performance	Some possible causes in case of low performance – *SAMPLE*	Corrective/preventive action
Check compressor full load current (Amps)	Match value on data plate	• Problem: high Amps • Effect: high energy consumption	• Low refrigerant level • Low motor efficiency	• Check and maintain refrigerant level • Check winding and restore
Measure variable frequency drive effectiveness	In-voltage is the same value as that of the main source voltage, out-voltage = % frequency	• Problem: higher/lower required power • Effect: higher energy consumption. Less efficiency	• Drive defected • Meters not calibrated • Ambient temperature • Load defected	• Check drive cooling fans and fix it • Maintain the ambient temperature to the accepted value • Check meters calibration and recalibrate • Check load parameter and adjust
Check condenser fan motor's full load current (Amps)	Match value on data plate	• Problem: high Amps • Effect: high energy consumption, low airflow rate extraction	• Blocked condenser coil • Worn bearing • Damaged winding	• Clean condenser coil • Replace bearing • Fix winding
Evaporator pressure drop (PSI)	Match value on data plate	• Problem: high-pressure drop • Effect: higher electrical consumption in pump's motor	• Scales accumulated inside the evaporator • Increased chilled water flow rate inside the coil than T&C	• Clean evaporator • Check DRV setting and adjust to limit flow as per T&C data
Refrigerant leaks test	No leak	• Problem: high Amps • Effect: high energy consumption, low airflow rate extraction	• Leakage in direction expansion cycle	• Detect leak and ensure direct expansion cycle has no leakage

Predictive maintenance

Maintenance task	Acceptable performance	Problem/effect in case of low performance	Some possible causes in case of low performance – *SAMPLE*	Corrective/preventive action
Check compressor motor's current (Amps)	Operating value within predefined alarm limits	• Problem: high Amps • Effect: high energy consumption	• Low refrigerant level • Low motor efficiency	• Check and maintain refrigerant level • Check winding and restore
Chilled water supply temperature (°C)	Operating value within predefined alarm limits	• Problem: high supply temperature • Effect: low cooling in facility	• Ineffective heat exchange on evaporator • Issues with expansion valve • Refrigerant leak • No sufficient heat exchange on consider	• Clean evaporator • Check and replace expansion valve based on enthalpy value • Detect leak and ensure direct expansion cycle has no leakage • Clean condenser coil and check condensing temperature and condenser effectiveness
Operating pressure (PSI)	Operating value within predefined alarm limits	• Problem: high-pressure drop • Effect: higher electrical consumption in pump's motor	• Scales accumulated inside the evaporator • Increased chilled water flow rate inside the coil than T&C	• Clean evaporator • Check DRV setting and adjust to limit flow as per T&C data

Table 9.10 Heat exchangers.

Equipment type: heat exchangers

Preventive maintenance

Maintenance inspection	Acceptable performance	Problem/effect in case of low performance	Some possible causes in case of low performance – *SAMPLE*	Corrective/preventive action
Measure pressure drop (PSI)	Match value on data plate	• Problem: high pressure drop • Effect: higher electrical consumption in pump's motor	• Scales accumulated inside the plates • Increased chilled water flow rate inside the plates than T&C	• Clean plates • Check DRV setting and adjust to limit flow as per T&C data
Heat exchanger effectiveness (%) – (in/out temperature)	Match value on data plate	• Problem: high water supply temperature • Effect: less cooling, depends on the function	• Scales accumulated inside the plates • Increased chilled water flow rate inside the plates than T&C • Higher in-water temperature than T&C values	• Clean plates • Check DRV setting and adjust to limit flow as per T&C data • Investigate reasons of high in-water temperature and rectify

Predictive maintenance

Maintenance task	Acceptable performance	Problem/effect in case of low performance	Some possible causes in case of low performance – *SAMPLE*	Corrective/preventive action
Chilled water supply temperature (°C)	Operating value within predefined alarm limits	• Problem: high water supply temperature • Effect: less cooling, depends on the function	• Scales accumulated inside the plates • Increased chilled water flow rate inside the plates than T&C • Higher in-water temperature than T&C values	• Clean plates • Check DRV setting and adjust to limit flow as per T&C data • Investigate reasons of high in-water temperature and rectify

Table 9.11 Water cooled chillers.

Equipment type: water cooled chillers

Preventive maintenance

Maintenance inspection	Acceptable performance	Problem/effect in case of low performance	Some possible causes in case of low performance – *SAMPLE*	Corrective/preventive action
Check compressor full load current (Amps)	Match value on data plate	• Problem: high Amps • Effect: high energy consumption	• Low refrigerant level • Low motor efficiency	• Check and maintain refrigerant level • Check winding and restore
Evaporator pressure drop (PSI)	Match value on data plate	• Problem: high pressure drop • Effect: higher electrical consumption in pump's motor	• Scales accumulated inside the evaporator • Increased chilled water flow rate inside the coil than T&C	• Clean evaporator • Check DRV setting and adjust to limit flow as per T&C data
Measure variable frequency drive effectiveness	In-voltage is the same value as that of main source voltage, out-voltage = % frequency	• Problem: higher/lower required power • Effect: higher energy consumption. Less efficiency	• Drive defected • Meters not calibrated • Ambient temperature • Load defected	• Check drive cooling fans and fix it • Maintain the ambient temperature to accepted value • Check meters calibration and recalibrate • Check load parameter and adjust
Condenser pressure drop (PSI)	Match value on data plate	• Problem: high pressure drop • Effect: higher electrical consumption in pump's motor	• Scales accumulated inside the condenser • Increased chilled water flow rate inside the coil than T&C	• Clean condenser • Check DRV setting and adjust to limit flow as per T&C data.
Refrigerant leaks test	No Leak	• Problem: high Amps • Effect: high energy consumption, low airflow rate extraction	• Leakage in direction expansion cycle	• Detect leak and ensure direct expansion cycle has no leakage

Table 9.11 Continued

Predictive maintenance				
Maintenance task	**Acceptable performance**	**Problem/effect in case of low performance**	**Some possible causes in case of low performance – *SAMPLE***	**Corrective/preventive action**
Check compressor motor's current (Amps)	Operating value within predefined alarm limits	• Problem: high Amps • Effect: high energy consumption	• Low refrigerant level • Low motor efficiency	• Check and maintain refrigerant level • Check winding and restore
Chilled water supply temperature (°C)	Operating value within predefined alarm limits	• Operating value within predefined alarm limits	• Problem: high supply temperature • Effect: low cooling in facility	• Ineffective heat exchange on evaporator • Issues with expansion valve • Refrigerant leak • No sufficient heat exchange on consider
Operating pressure (PSI)	Operating value within predefined alarm limits	• Operating value within predefined alarm limits	• Problem: high pressure drop • Effect: higher electrical consumption in pump's motor	• Scales accumulated inside the evaporator • Increased chilled water flow rate inside the coil than T&C

Table 9.12 Direct expansion air conditioners.

Equipment type: direct expansion air conditioners				
Preventive maintenance				
Maintenance inspection	**Acceptable performance**	**Problem/effect in case of low performance**	**Some possible causes in case of low performance – *SAMPLE***	**Corrective/preventive action**
Check compressor full load current (Amps)	Match value on data plate	• Problem: high Amps • Effect: high energy consumption	• Low refrigerant level • Low motor efficiency	• Check and maintain refrigerant level • Check winding and restore

Refrigerant leaks test	No leak	• Problem: high Amps • Effect: high energy consumption, low airflow rate extraction	• Leakage in direction expansion cycle	• Detect leak and ensure direct expansion cycle has no leakage

Predictive maintenance

Maintenance task	Acceptable performance	Problem/effect in case of low performance	Some possible causes in case of low performance – *SAMPLE*	Corrective/preventive action
Check compressor motor's current (Amps)	Operating value within predefined alarm limits	• Problem: high Amps • Effect: high energy consumption	• Low refrigerant level • Low motor efficiency	• Check and maintain refrigerant level • Check winding and restore

Table 9.13 Economizers.

Equipment type: economizers

Preventive maintenance

Maintenance inspection	Acceptable performance	Problem/effect in case of low performance	Some possible causes in case of low performance – *SAMPLE*	Corrective/preventive action
Supply air flow rate after mixing (m^3/hr)	Match value on data plate	• Problem: low airflow rate • Effect: depends on the application. If used for free cooling, it may result in low cooling inside the building	• Stuck dampers • Blocked air intakes	• Check auto function of dampers • Check air intakes for any blockage and clean
Supply air flow rate temperature after mixing (°C)	Match value on data plate	• Problem: high supply air temperature • Effect: depends on the application. If used for free cooling, it may result in low cooling inside the building	• Air mixing ratio not accurate, more return air is mixing with outside air	• Air flow rate requires re-balancing by adjusting dampers settings

Table 9.13 Continued

Predictive maintenance				
Maintenance task	**Acceptable performance**	**Problem/effect in case of low performance**	**Some possible causes in case of low performance – *SAMPLE***	**Corrective/preventive action**
Supply air flow rate after mixing (m^3/hr)	Match value on data plate	• Problem: low airflow rate • Effect: depends on the application. If used for free cooling, it may result in low cooling inside the building	• Stuck dampers • Blocked air intakes	• Check auto function of dampers • Check air intakes for any blockage and clean
Supply air flow rate temperature after mixing (°C)	Match value on data plate	• Problem: high supply air temperature • Effect: depends on the application. If used for free cooling, it may result in low cooling inside the building	• Air mixing ratio not accurate, more return air is mixing with outside air	• Air flow rate requires re-balancing by adjusting dampers settings

Table 9.14 Air compressors.

Equipment type: air compressors				
Preventive maintenance				
Maintenance inspection	**Acceptable performance**	**Problem/effect in case of low performance**	**Some possible causes in case of low performance – *SAMPLE***	**Corrective/preventive action**
Produced air pressure (PSI)	Match value on data plate	• Problem: low produced air pressure • Effect: impact on operational function	• Air leakage • Compressors degrading power	• Check and rectify if any air leakage is found • Check compressor effectiveness and rectify

9.5.2 Water Supply System

Table 9.15 Domestic water pump set, irrigation pump, and water features pumps.

Equipment type: pumps

Preventive maintenance

Maintenance inspection	Acceptable performance	Problem/effect in case of low performance	Some possible causes in case of low performance – *SAMPLE*	Corrective/preventive action
Measure full speed water flow rate (gpm, lps)	Match value on data plate	• Problem: low water flow rate • Effect: insufficient water delivery – based on function	• Clogged strainer • Air clogged in the system • Cavitation • Low motor speed • Torn impeller	• Clean strainer • Evacuate air from system • Fix motor winding
Check motor's full load current (Amps)	Match value on data plate	• Problem: low water flow • Effect: pump not capable of functioning as designed, high energy consumption	• Increase in the pump head • Worn bearing • Blocked strainers • Low motor efficiency	• Verify pump head and check impeller • Replace bearing • Clean strainer • Check winding and restore
Measure variable frequency drive effectiveness	In-voltage is the same value as that of the main source voltage, out-voltage = % frequency	• Problem: higher/lower required power • Effect: higher energy consumption. Less efficiency	• Drive defected • Meters not calibrated • Ambient temperature • Load defected	• Check drive cooling fans and fix it. • Maintain the ambient temperature to accepted value • Check meters calibration and recalibrate • Check load parameter and adjust

Table 9.15 Continued

Predictive maintenance				
Maintenance task	**Acceptable performance**	**Problem/effect in case of low performance**	**Some possible causes in case of low performance – *SAMPLE***	**Corrective/preventive action**
Measure water flow rate (gpm, lps)	Operating value within predefined alarm limits	• Problem: low water flow rate • Effect: insufficient water delivery – based on function	• Clogged strainer • Air clogged in the system • Cavitation • Low motor speed	• Clean strainer • Evacuate air from system • Fix motor winding
Measure pump's motor power consumption (kWh)	Operating value within predefined alarm limits	• Problem: kWh more than normal range • Effect: high energy consumption	• Damaged bearing • Blocked strainer • Damaged PF	• Replace bearing • Clean strainer • Replace PF
Motor running current	Operating value within predefined alarm limits	• Problem: kWh more than normal range • Effect: high energy consumption	• Damaged bearing • Blocked strainer • Damaged PF	• Replace bearing • Clean strainer • Replace PF

Table 9.16 Heat exchangers.

Equipment type: heat exchangers				
Preventive maintenance				
Maintenance inspection	**Acceptable performance**	**Problem/effect in case of low performance**	**Some possible causes in case of low performance – *SAMPLE***	**Corrective/preventive action**
Measure pressure drop (PSI)	Match value on data plate	• Problem: high pressure drop • Effect: higher electrical consumption in pump's motor	• Scales accumulated inside the plates • Increased chilled water flow rate inside the plates than T&C	• Clean plates • Check DRV setting and adjust to limit flow as per T&C data
Heat exchanger effectiveness (%) – (in/ out temperature)	Match value on data plate	• Problem: high water supply temperature • Effect: less cooling, depends on the function	• Scales accumulated inside the plates • Increased chilled water flow rate inside the plates than T&C • Higher in-water temperature than T&C values	• Clean plates • Check DRV setting and adjust to limit flow as per T&C data • Investigate reasons of high in-water temperature and rectify
Predictive maintenance				
Maintenance task	**Acceptable performance**	**Problem/effect in case of low performance**	**Some possible causes in case of low performance – *SAMPLE***	**Corrective/preventive action**
Chilled water supply temperature (°C)	Operating value within predefined alarm limits	• Problem: high water supply temperature • Effect: less cooling, depends on the function	• Scales accumulated inside the plates • Increased chilled water flow rate inside the plates than T&C • Higher in-water temperature than T&C values	• Clean plates • Check DRV setting and adjust to limit flow as per T&C data • Investigate reasons of high in-water temperature and rectify

Table 9.17 Pressure reducing valve station.

Equipment type: PRV station

Preventive maintenance

Maintenance inspection	**Acceptable performance**	**Problem/effect in case of low performance**	**Some possible causes in case of low performance – *SAMPLE***	**Corrective/preventive action**
Measure on/off pressure on critical PRVs	Achieve pressure value as per T&C	Problem: higher (or lower) water pressure than T&C Effect: increase in-water consumption in case of higher water pressure, or dissatisfaction of building occupants in case of low water pressure	Damaged PRV or change in settings during its operating life Dirty screen inside PRV leads to low outlet pressure	Check settings and recalibrate when required Clean screen and recalibrate

Note: Only main PRVs should be recalibrated. For example, recalibrate PRVs mainly on domestic water lines entering floors.

Table 9.18 Boilers.

Equipment type: boilers				
Preventive maintenance				
Maintenance inspection	**Acceptable performance**	**Problem/effect in case of low performance**	**Some possible causes in case of low performance – *SAMPLE***	**Corrective/preventive action**
Measure fuel combustion efficiency	±5% of original efficiency	• Problem: low efficiency • Effect: insufficient supply water temperature/high energy consumption	• Air fuel ratio is not balanced • Combustion process is not efficient • Supply of high amount of water, more than design flow rate	• Test and calibrate air/fuel ratio • Do all required cleaning at burners, heat exchangers, etc. • Supply water flow rate as per design and check on regular basis
Inspect steam leakage	No steam leakage	• Problem: steam leakage • Effect: high energy consumption	• Leakage in tubes, plates, etc.	• Identify leakage location and restore • Check pressure on regular basis
Predictive maintenance				
Maintenance task	**Acceptable performance**	**Problem/effect in case of low performance**	**Some possible causes in case of low performance – *SAMPLE***	**Corrective/preventive action**
Outlet water temperature	Operating value within predefined alarm limits	• Problem: low water temperature • Effect: insufficient supply water temperature/high energy consumption	• Air fuel ratio is not balanced • Combustion process is not efficient • Supply of high amount of water, more than design flow rate	• Test and calibrate air/fuel ratio • Do all required cleaning at burners, heat exchangers, etc. • Supply water flow rate as per design and check on a regular basis
Primary system water pressure	Operating value within predefined alarm limits	• Problem: steam leakage • Effect: high energy consumption	• Leakage in tubes, plates, etc.	• Identify leakage location and restore • Check pressure on regular basis

9.5.3 Drainage System

Table 9.19 Sump pumps.

Equipment type: pumps

Preventive maintenance

Maintenance inspection	Acceptable performance	Problem/effect in case of low performance	Some possible causes in case of low performance – *SAMPLE*	Corrective/preventive action
Measure full speed water flow rate (gpm, lps)	Match value on data plate	• Problem: low water flow rate • Effect: insufficient water delivery – based on function	• Clogged strainer • Air clogged in the system • Cavitation • Low motor speed • Torn impeller	• Clean strainer • Evacuate air from system • Fix motor winding
Check motor's full load current (Amps)	Match value on data plate	• Problem: low water flow • Effect: pump not capable of functioning as designed, high energy consumption	• Increase in the pump head • Worn bearing • Blocked strainers • Low motor efficiency	• Verify pump head and check impeller • Replace bearing • Clean strainer • Check winding and restore
Measure variable frequency drive effectiveness	In-voltage is the same value as that of main source voltage, out-voltage = % frequency	• Problem: higher/lower required power • Effect: higher energy consumption. Less efficiency	• Drive defected • Meters not calibrated • Ambient temperature • Load defected	• Check drive cooling fans and fix it • Maintain the ambient temperature to accepted value • Check meters calibration and recalibrate • Check load parameter and adjust

Predictive maintenance

Maintenance task	Acceptable performance	Problem/effect in case of low performance	Some possible causes in case of low performance – *SAMPLE*	Corrective/preventive action
Measure water flow rate (gpm, lps)	Operating value within predefined alarm limits	• Problem: low water flow rate • Effect: insufficient water delivery – based on function	• Clogged strainer • Air clogged in the system • Cavitation • Low motor speed	• Clean strainer • Evacuate air from system • Fix motor winding

Measure pump's motor power consumption (kWh)	Operating value within predefined alarm limits	• Problem: kWh more than normal range • Effect: high energy consumption	• Damaged bearing • Blocked strainer • Damaged PF	• Replace bearing • Clean strainer • Replace PF
Motor running current	Operating value within predefined alarm limits	• Problem: kWh more than normal range • Effect: high energy consumption	• Damaged bearing • Blocked strainer • Damaged PF	• Replace bearing • Clean strainer • Replace PF

9.5.4 Storm Water Management System

Table 9.20 Rainwater pumps.

Equipment type: pumps

Preventive maintenance

Maintenance inspection	Acceptable performance	Problem/effect in case of low performance	Some possible causes in case of low performance – *SAMPLE*	Corrective/preventive action
Measure full speed water flow rate (gpm, lps)	Match value on data plate	• Problem: low water flow rate • Effect: insufficient water delivery – based on function	• Clogged strainer • Air clogged in the system • Cavitation • Low motor speed • Torn impeller	• Clean strainer • Evacuate air from system • Fix motor winding
Check motor's full load current (Amps)	Match value on data plate	• Problem: low water flow • Effect: pump not capable of functioning as designed, high energy consumption	• Increase in the pump head. • Worn bearing • Blocked strainers • Low motor efficiency	• Verify pump head and check impeller • Replace bearing • Clean strainer • Check winding and restore
Measure variable frequency drive effectiveness	In-voltage is the same value as that of the main source voltage, out-voltage = % frequency	• Problem: higher/lower required power • Effect: higher energy consumption. Less efficiency	• Drive defected • Meters not calibrated • Ambient temperature • Load defected	• Check drive cooling fans and fix it • Maintain the ambient temperature to accepted value • Check meters calibration and recalibrate • Check load parameter and adjust

Table 9.20 Continued

Predictive maintenance				
Maintenance task	**Acceptable performance**	**Problem/effect in case of low performance**	**Some possible causes in case of low performance – *SAMPLE***	**Corrective/preventive action**
Measure water flow rate (gpm, lps)	Operating value within predefined alarm limits	• Problem: low water flow rate • Effect: insufficient water delivery – based on function	• Clogged strainer • Air clogged in the system • Cavitation • Low motor speed	• Clean strainer • Evacuate air from system • Fix motor winding
Measure pump's motor power consumption (kWh)	Operating value within predefined alarm limits	• Problem: kWh more than normal range • Effect: high energy consumption	• Damaged bearing • Blocked strainer • Damaged PF	• Replace bearing • Clean strainer • Replace PF
Motor running current	Operating value within predefined alarm limits	• Problem: kWh more than normal range • Effect: high energy consumption	• Damaged bearing • Blocked strainer • Damaged PF	• Replace bearing • Clean strainer • Replace PF

9.5.5 Building Transportation System

Table 9.21 Travelators and escalators.

Equipment type: travelators and escalators

Preventive maintenance

Maintenance inspection	Acceptable performance	Problem/effect in case of low performance	Some possible causes in case of low performance – *SAMPLE*	Corrective/preventive action
Check motor's full load current (Amps)	Match value on data plate	• Problem: high Amps • Effect: higher energy consumption	• Low motor efficiency • Damaged winding • High operating temperature	• Check motor efficiency and repair • Repair or replace motor winding • Maintain acceptable temperature in surrounding environment
Auto start/stop command	Auto start/stop to be workable	• Problem: motor continuously working • Effect: higher energy consumption	• Damage on/off sensors • Control logic stuck on (on mode)	• Check functionality of sensor and replace defected ones • Check control logic for auto start/stop command

Predictive maintenance

Maintenance task	Acceptable performance	Problem/effect in case of low performance	Some possible causes in case of low performance – *SAMPLE*	Corrective/preventive action
Measure motor power consumption (kWh)	Operating value within predefined alarm limits	• Problem: high Amps • Effect: higher energy consumption	• Low motor efficiency • Damaged winding • High operating temperature	• Check motor efficiency and repair • Repair or replace motor winding • Maintain acceptable temperature in surrounding environment

Table 9.22 Elevators.

Equipment type: elevators

Preventive maintenance

Maintenance inspection	**Acceptable performance**	**Problem/effect in case of low performance**	**Some possible causes in case of low performance – *SAMPLE***	**Corrective/preventive action**
Check motor's full load current (Amps)	Match value on data plate	• Problem: high Amps • Effect: higher energy consumption	• Low motor efficiency • Damaged winding • High operating temperature	• Check motor efficiency and repair • Repair or replace motor winding • Maintain acceptable temperature in surrounding environment
Auto start/stop command	Auto start/stop to be workable	• Problem: motor continuously working • Effect: higher energy consumption	• Damage on/off sensors • Control logic stuck on (on mode)	• Check functionality of sensor and replace defected ones • Check control logic for auto start/stop command

Predictive maintenance

Maintenance task	**Acceptable performance**	**Problem/effect in case of low performance**	**Some possible causes in case of low performance – *SAMPLE***	**Corrective/preventive action**
Measure motor power consumption (kWh)	Operating value within predefined alarm limits	• Problem: high Amps • Effect: higher energy consumption	• Low motor efficiency • Damaged winding • High operating temperature	• Check motor efficiency and repair • Repair or replace motor winding • Maintain acceptable temperature in surrounding environment

9.5.6 Fire Fighting System

Table 9.23 Fire pumps.

Equipment type: pumps
Preventive maintenance

Maintenance inspection	Acceptable performance	Problem/effect in case of low performance	Some possible causes in case of low performance – *SAMPLE*	Corrective/preventive action
Measure full speed water flow rate (gpm, lps)	Match value on data plate	• Problem: low water flow rate. • Effect: insufficient water delivery – based on function	• Clogged strainer • Air clogged in the system • Cavitation • Low motor speed • Torn impeller	• Clean strainer • Evacuate air from system • Fix motor winding
Check motor's full load current (Amps)	Match value on data plate	• Problem: low water flow • Effect: pump not capable of functioning as designed, high energy consumption	• Increase in the pump head • Worn bearing • Blocked strainers • Low motor efficiency	• Verify pump head and check impeller • Replace bearing • Clean strainer • Check winding and restore
Measure variable frequency drive effectiveness	In-voltage is the same value as that of main source voltage, out-voltage = % frequency	• Problem: higher/lower required power • Effect: higher energy consumption. Less efficiency	• Drive defected • Meters not calibrated • Ambient temperature • Load defected	• Check drive cooling fans and fix it • Maintain the ambient temperature to accepted value • Check meters calibration and recalibrate • Check load parameter and adjust

Table 9.23 Continued

Predictive maintenance				
Maintenance task	**Acceptable performance**	**Problem/effect in case of low performance**	**Some possible causes in case of low performance – *SAMPLE***	**Corrective/preventive action**
Measure water flow rate (gpm, lps)	Operating value within predefined alarm limits	• Problem: low water flow rate. • Effect: insufficient water delivery – based on function	• Clogged strainer • Air clogged in the system • Cavitation • Low motor speed	• Clean strainer • Evacuate air from system • Fix motor winding
Measure pump's motor power consumption (kWh)	Operating value within predefined alarm limits	• Problem: kWh more than normal range • Effect: high energy consumption	• Damaged bearing • Blocked strainer • Damaged PF	• Replace bearing • Clean strainer • Replace PF
Motor running current	Operating value within predefined alarm limits	• Problem: kWh more than normal range • Effect: high energy consumption	• Damaged bearing • Blocked strainer • Damaged PF	• Replace bearing • Clean strainer • Replace PF

9.5.7 Electrical System

Table 9.24 Motor control center.

Equipment type: MCC – starters

Preventive maintenance

Maintenance inspection	Acceptable performance	Problem/effect in case of low performance	Some possible causes in case of low performance – *SAMPLE*	Corrective/preventive action
In/out voltage at constant frequency	In-voltage is the same value as that of main source voltage, out-voltage = % frequency	• Problem: higher/lower required power • Effect: higher energy consumption. Less efficiency	• Drive defected • Meters not calibrated • Ambient temperature • Load defected	• Check drive cooling fans and fix it • Maintain the ambient temperature to accepted value • Check meters calibration and recalibrate • Check load parameter and adjust
In/out current at constant frequency	In-current = out current +2%	• Problem: higher/lower required power • Effect: higher energy consumption. Less efficiency	• Drive defected • Meters not calibrated • Ambient temperature • Load defected	• Check drive cooling fans and fix it • Maintain the ambient temperature to accepted value • Check meters calibration and recalibrate • Check load parameter and adjust
Operating temperature	In accordance with manufacturer recommendations	• Problem: operating temperature increase lead to drive defect • Effect: more energy consumption	• Drive defected	• Maintain the ambient temperature to accepted value

Predictive maintenance

Maintenance task	Acceptable performance	Problem/effect in case of low performance	Some possible causes in case of low performance – *SAMPLE*	Corrective/preventive action
Verify the out-voltage with frequency	In-voltage is the same value as that of main source voltage, out-voltage = % frequency	• Problem: higher/lower required power • Effect: higher energy consumption. Less efficiency	• Drive defected • Meters not calibrated • Ambient temperature • Load defected	• Check drive cooling fans and fix it • Maintain the ambient temperature to accepted value • Check meters calibration and recalibrate • Check load parameter and adjust
Ambient temperature	In accordance with manufacturer recommendations	• Problem: operating temperature increase lead to drive defect • Effect: more energy consumption	• Drive defected	• Maintain the ambient temperature to accepted value

Table 9.25 Variable frequency drive.

Equipment type: VFD

Preventive maintenance

Maintenance inspection	**Acceptable performance**	**Problem/effect in case of low performance**	**Some possible causes in case of low performance – *SAMPLE***	**Corrective/preventive action**
In/out voltage at constant frequency	In-voltage is the same value as that of the main source voltage, out-voltage = % frequency	• Problem: higher/lower required power • Effect: higher energy consumption. Less efficiency	• Drive defected • Meters not calibrated • Ambient temperature • Load defected	• Check drive cooling fans and fix it • Maintain the ambient temperature to accepted value • Check meters calibration and recalibrate • Check load parameter and adjust
In/out current at constant frequency	In-current = out current +2%	• Problem: higher/lower required power • Effect: higher energy consumption. Less efficiency	• Drive defected • Meters not calibrated • Ambient temperature • Load defected	• Check drive cooling fans and fix it • Maintain the ambient temperature to accepted value • Check meters calibration and recalibrate • Check load parameter and adjust
Operating temperature	In accordance with manufacturer recommendations	• Problem: operating temperature increase lead to drive defect • Effect: more energy consumption	• Drive defected	• Maintain the ambient temperature to accepted value

Predictive maintenance

Maintenance task	**Acceptable performance**	**Problem/effect in case of low performance**	**Some possible causes in case of low performance – *SAMPLE***	**Corrective/preventive action**
Verify the out-voltage with frequency	In-voltage is the same value as that of the main source voltage, out-voltage = % frequency	• Problem: higher/lower required power • Effect: higher energy consumption. Less efficiency	• Drive defected • Meters not calibrated • Ambient temperature • Load defected	• Check drive cooling fans and fix it • Maintain the ambient temperature to accepted value • Check meters calibration and recalibrate • Check load parameter and adjust
Ambient temperature	In accordance with manufacturer recommendations	• Problem: operating temperature increase lead to drive defect • Effect: more energy consumption	• Drive defected	• Maintain the ambient temperature to accepted value

Table 9.26 Lighting bulbs.

Equipment type: PRV station

Preventive maintenance

Maintenance inspection	Acceptable performance	Problem/effect in case of low performance	Some possible causes in case of low performance – *SAMPLE*	Corrective/preventive action
Measure lux level (foot-candle)	Match value on manufacture catalogue	• Problem: low lux level • Effect: low illumination	• Dirty light bulb	• Clean bulbs on regular basis

9.5.8 Building Management System

Table 9.27 Two-way control valve.

Equipment Type: Two-way control valve

Preventive maintenance

Maintenance inspection	Acceptable performance	Problem/effect in case of low performance	Some possible causes in case of low performance – *SAMPLE*	Corrective/preventive action
Check actuator response in line with control signal	Linear response to control signal on site (0–10 V_{dc} = 0%–100% valve closing/opening)	• Problem: valve stuck and not regulating according to demand • Effect: overcooling/higher chilled water consumption	• Mechanical fault, valve stuck on open position • Lost signal from BMS • Faulty logic	• Check valve position and return to auto mode • Check valve response to BMS signal on regular basis • Check valve response to control logic
Check feedback correspondence to control signal	Response received on frontend software	• Problem: command is not received on-site equipment • Effect: overcooling/higher chilled water consumption	• Mechanical fault, valve stuck on open position • Lost signal from BMS • Faulty logic	• Check valve position and return to auto mode • Check valve response to BMS signal on regular basis • Check valve response to control logic

Table 9.27 Continued

Equipment Type: Two-way control valve

Preventive maintenance

Maintenance inspection	Acceptable performance	Problem/effect in case of low performance	Some possible causes in case of low performance – *SAMPLE*	Corrective/preventive action
Check response time with respect to command	Closing/opening time as per original logic (response time may be adjusted for more efficient response)	• Problem: valve closing/ opening response is slow • Effect: overcooling or/ higher chilled water consumption • Effect: insufficient cooling/ inconvenient space temperature and humidity	• Faulty logic	• Correct logic and check opening/ closing speed

Predictive maintenance

Maintenance task	Acceptable performance	Problem/effect in case of low performance	Some possible causes in case of low performance – *SAMPLE*	Corrective/preventive action
Valves response to AHU/FCU temperature set point	Linear response to set points	• Problem: valve closing/ opening response is slow • Effect: overcooling or/ Higher chilled water consumption • Effect: insufficient cooling/ inconvenient space temperature and humidity	• Faulty logic	• Correct logic and check opening/ closing speed

Table 9.28 Differential pressure switch.

Equipment type: Differential pressure switch

Preventive maintenance

Maintenance inspection	**Acceptable performance**	**Problem/effect in case of low performance**	**Some possible causes in case of low performance – *SAMPLE***	**Corrective/preventive action**
Check DPS functionality with respect to control signal	Linear response to control signal on site/true signal based on site condition in fans.	• Problem: fan's speed is not responding to DPS value • Effect: higher energy consumption	• Mechanical fault, fan operating on manual speed • Lost signal from BMS • Faulty logic	• Return fan to auto mode • Check fan speed response to DPS value and control logic on regular basis
Conduct continuity loop testing	Command/response received on frontend software and on-site equipment	• Problem: command is not received on on-site equipment • Effect: higher energy consumption	• Mechanical fault, fan operating on manual speed • Lost signal from BMS • Faulty logic	• Return fan to auto mode • Check fan speed response to DPS value and control logic on a regular basis
Compare differential pressure value on frontend and on-site	Command/response received on frontend software and on-site equipment	• Problem: different DPS reading on frontend and on-site • Effect: higher energy consumption	• Faulty DPS	• Replace DPS or recalibrate if possible

Predictive maintenance

Maintenance task	**Acceptable performance**	**Problem/effect in case of low performance**	**Some possible causes in case of low performance – *SAMPLE***	**Corrective/preventive action**
Differential pressure value	Value within predefined limits	• Problem: DPS reading outside acceptable limits • Effect: higher energy consumption	• Faulty DPS • Faulty logic resulted in higher DPS reading	• Replace DPS or recalibrate if possible. • Conduct continuity loop testing and rectify logic

Table 9.29 Differential pressure transmitter.

Equipment type: differential pressure transmitter

Preventive maintenance

Maintenance inspection	Acceptable performance	Problem/effect in case of low performance	Some possible causes in case of low performance – *SAMPLE*	Corrective/preventive action
Check DPT functionality with respect to control signal	Linear response to control signal on site/true signal based on site condition	• Problem: pump speed is not responding to DPT value • Effect: higher energy consumption	• Mechanical fault, fan operating on manual speed • Lost signal from BMS • Faulty logic	• Return pump to auto mode • Check pump speed response to DPT value and control logic on regular basis
Conduct continuity loop testing	Command/response received on frontend software and on-site equipment	• Problem: command is not received on on-site equipment • Effect: higher energy consumption	• Mechanical fault, pump operating on manual speed • Lost signal from BMS • Faulty logic	• Return pump to auto mode • Check pump speed response to DPT value and control logic on regular basis
Compare differential pressure value on frontend and on-site	Command/response received on frontend software and on-site equipment	• Problem: different DPT reading on frontend and on-site • Effect: higher energy consumption	• Faulty DPT	• Replace DPT or recalibrate if possible

Predictive maintenance

Maintenance task	Acceptable performance	Problem/effect in case of low performance	Some possible causes in case of low performance – *SAMPLE*	Corrective/preventive action
Differential pressure value	Value within predefined limits	• Problem: DPT reading outside acceptable limits • Effect: higher energy consumption	• Faulty DPT • Faulty logic resulted in higher DPT reading • Pump stuck on constant speed and not responding to DPT set point	• Replace DPT, or recalibrate if possible • Conduct continuity loop testing and rectify logic • Check pump VFD and respond to control logic and rectify

Table 9.30 Flow rate/velocity meters.

Equipment type: flow rate/velocity meters

Preventive maintenance

Maintenance inspection	**Acceptable performance**	**Problem/effect in case of low performance**	**Some possible causes in case of low performance – *SAMPLE***	**Corrective/preventive action**
Compare flow rate value on frontend and actual on-site value	98%–100% match between actual site reading and frontend value	• Problem: flow rate value on-site is higher/lower than the frontend • Effect: if higher flow rate, higher energy consumption • Effect: if lower flow rate: the machine is not achieving intended function (i.e., AHU is not delivering required flow rate)	• Not calibrated flowmeter • Fault reading in BMS	• Calibrate meter on a regular basis • Check reading on a regular basis
Conduct continuity loop testing	Command/response received on frontend software and on-site device	• Problem: command is not received on on-site equipment • Effect: possible higher energy consumption	• Lost signal from BMS • Damaged cables • Damaged BMS ports	• Check end-to-end BMS signal • Replace damaged cables • Replace damaged ports
Check flow rate response to software command (increase/decrease)	Command/response received on frontend software and on-site device	• Problem: motor (fan or pump) is not responding to BMS command • Effect: possible higher energy consumption, or lower delivered flow rate which is preventing the equipment from achieving its intended function	• Motor (fan/pump) operating on manual • Motor is on auto, but BMS signal is lost	• Operate motor on auto • Check end-to-end BMS signal and rectify on a regular basis

Predictive maintenance

Maintenance task	**Acceptable performance**	**Problem/effect in case of low performance**	**Some possible causes in case of low performance – *SAMPLE***	**Corrective/preventive action**
Flow rate/velocity value	Value within predefined limits	• Site is higher/lower than predefined limits • Effect: if higher flow rate, higher energy consumption • Effect: if lower flow rate: the machine is not achieving intended function	• Causes may be related to the equipment itself, BMS command, mechanical faults, etc.	• Case-by-case scenario, identify corrective/preventive actions and implement on a regular basis

Table 9.31 Cooling coil temperature and humidity sensors.

Equipment type: cooling coil temperature and humidity sensors				
Preventive maintenance				
Maintenance inspection	**Acceptable performance**	**Problem/effect in case of low performance**	**Some possible causes in case of low performance – *SAMPLE***	**Corrective/preventive action**
Check the sensor ohmic value and actual temperature where it is installed	98%–100% match between actual site reading and frontend value	• Problem: actual temperature is lower/higher than BMS value • Effect: if lower temperature, then higher energy consumption is recorded • Effect: if higher temperature, then insufficient cooling is supplied to the room	• Dirty sensor • Faulty sensor	• Clean sensor • Replace sensor
Predictive maintenance				
Maintenance inspection	**Acceptable performance**	**Problem/effect in case of low performance**	**Some possible causes in case of low performance – *SAMPLE***	**Corrective/preventive action**
Frontend recorded temperature value	Value within predefined limits	• Problem: temperature is correct but lower/higher than set limits • Effect: if lower temperature, then higher energy consumption is recorded • Effect: if higher temperature, then insufficient cooling is supplied to the room	• Two-way valve is not responding to set point (refer to two-way valve section)	• Refer to two-way valve section

Table 9.32 Chilled water temperature sensors.

Equipment type: chilled water temperature sensors

Preventive maintenance

Maintenance inspection	Acceptable performance	Problem/effect in case of low performance	Some possible causes in case of low performance – *SAMPLE*	Corrective/preventive action
Check the sensor ohmic value and actual temperature where it is installed	98%–100% match between actual site reading and frontend value	• Problem: actual temperature is lower/higher than BMS value • Effect: if lower temperature, then higher energy consumption is recorded • Effect: if higher temperature, then insufficient water temperature supplied to the equipment (i.e., AHU)	• Dirty sensor • Faulty sensor	• Clean sensor • Replace sensor

Predictive maintenance

Maintenance inspection	Acceptable performance	Problem/effect in case of low performance	Some possible causes in case of low performance – *SAMPLE*	Corrective/preventive action
Frontend recorded temperature value	Value within predefined limits	• Problem: temperature is correct but lower/higher than set limits • Effect: if lower temperature, then higher energy consumption is recorded • Effect: if higher temperature, then insufficient water temperature supplied to the equipment (i.e., AHU)	• Issues with source of water (refer to chillers and heat exchangers sections)	• Refer to chillers and heat exchangers sections

Table 9.33 Space/return air temperature and humidity sensors.

Equipment type: space/return air temperature and humidity sensors

Preventive maintenance

Maintenance Inspection	Acceptable Performance	Problem/effect in case of low performance	Some possible causes in case of low performance – *SAMPLE*	Corrective/preventive action
Check the sensor ohmic value and actual temperature where it is installed	98%–100% match between actual site reading and frontend value	• Problem: actual space temperature is lower/higher than BMS value or set point • Effect: if lower temperature, then higher energy consumption is recorded • Effect: if higher temperature, then insufficient cooling resulted in the space	• Faulty space temperature sensor • Wrong BMS reading	• Replace sensor • Check and correct BMS reading according to actual reading
Check response of two-way valve to space/ return air temperature	Two-way valve open/close as per control logic	• Refer to two-way valve section	• Refer to two-way valve section	• Refer to two-way valve section

Predictive maintenance

Maintenance task	Acceptable performance	Problem/effect in case of low performance	Some possible causes in case of low performance – *SAMPLE*	Corrective/preventive action
Frontend recorded temperature value	Value within predefined limits	• Problem: temperature is lower/higher than set limits • Effect: if lower temperature, then higher energy consumption is recorded • Effect: if higher temperature, then insufficient cooling resulted in the space	• Different possible causes could be in relation with mechanical equipment or control logic as described in previous sections	• Different possible corrective/ preventive actions related to mechanical equipment or control logic as described in previous sections

Table 9.34 Control logic for all equipment controlled by BMS.

Equipment type: control logic for all equipment controlled by BMS

Preventive maintenance

Maintenance inspection	**Acceptable performance**	**Problem/effect in case of low performance**	**Some possible causes in case of low performance – *SAMPLE***	**Corrective/preventive action**
Control logic for all equipment that is part of energy centered maintenance process, for example: • AHUs • FCUs • FAHUs • Heat recovery units • Ecology units • Close control units • Pumps • Swimming pools cooling system • Chillers • Heat exchangers • Fans • Lifts and escalators • Travelators • Boilers • Cooling towers • DX units • MCC • VFDs	Control logic to be in accordance with system design sequence of operation	• Case-by-case scenario; refer to previous sections	• Case-by-case scenario; refer to previous sections	• Case-by-case scenario; refer to previous sections

10

Energy Centered Maintenance to avoid Low Delta T Syndrome in Chilled Water Systems

This chapter is tackling another important topic that impacts the energy efficiency of equipment due to lack of proper maintenance and due to inefficient control logics, which is about the impact of low return chilled water temperature (so-called delta T syndrome) on chilled water system energy efficiency.

This chapter provides the basic definition of low delta T syndrome and investigates some of the possible causes of it that result from lack of maintenance and the ways of mitigating this problem.

This chapter describes what kind of maintenance tasks should be implemented during the planned preventive maintenance program to reduce or eliminate the causes of low delta T in chilled water systems. The maintenance tasks will be described in a similar manner as in previous chapters as part of energy centered maintenance model.

Effective maintenance measures and tasks can limit or reduce the impact of low delta T on equipment's energy consumption, system effectiveness, and thermal comfort level in the buildings, thus saving energy and avoiding additional charges and costs.

10.1 Low Delta T Syndrome Described

Low delta T syndrome is almost a common problem in variable flow chilled water systems; low delta T is defined such that the actual difference between supply and return chilled water temperatures (delta T) is less than the design value. In a simpler definition, low delta T occurs when the return chilled water temperature from the building is lower than the design returns chilled water temperature, which is the case in most buildings operating with variable flow chilled water system. A typical Schematic layout of chilled water loop is shown in figure 10.1.

Low delta T results in reducing the effectiveness of chillers and increases the energy consumption of the cooling plant. Lower return

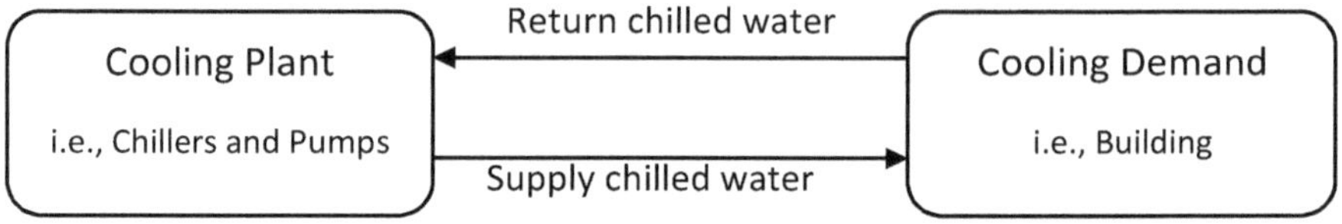

Figure 10.1 Schematic layout of supply/return chilled water loop.

chilled water temperature decreases the temperature difference (ΔT) across the chiller as the water supply temperature is fixed to a set point; therefore, the required chilled water flow rate will be larger to meet the expected cooling demand. The cooling plant performance is significantly reduced by lower operating chilled water delta T than design conditions, and pumping energies are significantly increased to supply the increase in chilled water flow rate.

The following heat transfer equation represents the relation between the chilled water flow rate and chilled water temperature difference (delta T):

$$Q = m\ C_p \Delta T \ldots (10.1)$$

where:
Q is the heat transfer, referred to as cooling load in kJ;
$\dot{m}$ is the water mass flow rate in m^3/s;

C_p is the specific heat of water in kJ/kg.K;
ΔT is the temperature difference in Kelvin (K).

The above equation shows that the load is directly proportional to mass flow rate and temperature reference. Therefore, if the delta T remains constant, the required mass flow rate will remain constant to give the same load, while if the temperature difference drops, the mass flow rate should increase to meet the same load.

The same principle is happening in the chilled water system that is suffering from low delta T syndrome, where the system fails to maintain a constant delta T.

In this case, to keep the same heat transfer (cooling load generated by the chillers), the mass flow (chilled water flow rate) should increase in case the delta T (temperature difference between supply and return chilled water temperatures) drops. The increase in chilled water flow rate will increase the pumping energy of chilled water pumps, and it will reduce the chiller effectiveness which will result in adding more chillers into operation to meet the required chilled water flow rate and cooling demand of the building.

Chiller effectiveness is calculated based on the ratio between actual delta T value to design delta T value as shown in the following formula:

$$Chiller\ Effectivnes = \left(\frac{ARCWT - ASCWT}{DRCWT - DSCWT}\right) \times 100\% \text{ ......(10.2)}$$

- Actual return chilled water temperature (ºC) is referred to as ARCWT.
- Actual supply chilled water temperature (ºC) is referred to as ASCWT.
- Design return chilled water temperature (ºC) is referred to as DRCWT.
- Design supply chilled water temperature (ºC) is referred to as DSCWT.

If the actual chilled water temperature drops by a certain value from the design chilled water temperature, it will have a direct impact on the chiller's effectiveness. For example, if the required delta T according to design is 8.0 ºC, while the actual delta T is 6.4 ºC, the chiller effectiveness drops from 100% to 80%. In such a case, the same existing chillers cannot meet the building demand because the effectiveness has dropped, which will require additional chillers to be put into operation to meet the required demand. It is important to highlight that the actual number of chillers needed to operate at part load operation is decided based on chiller capacity and based on the control logic of the cooling plant.

Low delta T has a direct impact on energy consumption of the facility and effectiveness of the chilled water system. As explained earlier, when the delta T drops, it increases the energy consumption of the chilled water pumps and chillers by increasing the pumping energy and by turning on more chillers and pumps into operation to meet the same cooling demand.

Low delta T syndrome results from the inefficient use of chilled water amount at the building and plant side, as well as due to lack of maintenance of chilled water system components which is yielding a lower return chilled water temperature than the design value. Maintaining the right delta T in chilled water systems increases the energy efficiency of the cooling plant and reduces potential energy waste during plant operation.

10.2 Maintenance Relationship

Lower delta T syndrome results from the inefficient use of chilled water at the building and plant side, yielding a lower chilled water return temperature than the design value. There are multiple causes of low delta T which could be related to design issues, installation issues, and, most importantly, due to improper operation and maintenance.

Energy centered maintenance model focuses on the issues appearing during the operation of the equipment and provides guidance on identifying the root cause of the problem and the way to mitigate by conducting certain corrective and preventive actions.

This section identifies some of the possible causes of low delta T that are related to lack of maintenance. However, multiple types of research and studies have been conducted by other researchers who identified the possible causes of the problem from the design perspective; some of those causes will also be listed for the benefit of the reader.

Steven T. Taylor has issued a paper titled "Degrading Chilled Water Plant Delta-T: Causes and Mitigations" in which he listed most possible causes of the problem and the way to mitigate it. Another paper by Donald P. Fiorino titled "Achieving High Chilled Water Delta Ts" also discusses 25 best practices to achieve high chilled water temperature difference. Some of the points considered by both authors are discussed in this chapter.

10.3 Causes Can Be Avoided During Design Stage

Some of low Delta causes can be due to either design faults related to the selection of an improper type of equipment or poor control logic of chilled water system components. Those causes can be eliminated by taking into consideration the right factors in new designs which can mitigate this problem.

10.3.1 Use of Constant Flow Chilled Water System

The use of constant flow chilled water system maintains the same supply chilled water flow rate amount to fan coil units (FCUs) until the set point temperature inside the room is achieved. Till then, the chilled water temperature varies with the load, and it could drop below design return temperature and causes low delta T. Corrective/Preventative action for this cause is mentioned in table 10.1:

Table 10.1 Use of constant flow chilled water system.

Cause	Corrective/preventive action
Use of constant flow chilled water system	Use variable flow chilled water system instead of constant flow along with selecting proper equipment size, control valve type and size, and control logic

10.3.2 Use of Three-Way Control Valve in Variable Flow Chilled Water System

Most control valves used in variable flow chilled water system are two-way control valves. However, designers tend to keep few three-way valves in the system to allow chilled water flow rate to circulate. Three-way valves are not connected to the load (not to a fan coil unit). Instead, they may be provided at the end of the chilled water network for water circulation. In this case, the same supply chilled water temperature returns to the cooling plant, which affects the total return chilled water temperature from the building and could lead to low delta T. Corrective/Preventative action for this cause is mentioned in table 10.2:

Table 10.2 Use of three-way control valve in variable flow chilled water system.

Cause	Corrective/preventive action
Use of three-way control valve in variable flow chilled water system	Do not use three-way valves in variable flow systems for chilled water circulating. The intent of circulating the water is to reduce the time the chilled water will reach to an FCU when it requires it. The reality is that the chilled water flow is always circulating in the system even in low demand time

10.3.3 Cooling Coil Selection for Low Delta T than Design

Sometimes fan coil units cooling coils are selected for lower supply/return temperature difference than design. In this case, and even if the units are functioning its best performance, the return chilled water temperature will always be less than the required temperature just because the coil is designed for lower temperature difference and it will cause a low delta T. Corrective/Preventative action for this cause is mentioned in table 10.3:

Table 10.3 Cooling coil selection for low delta T than design.

Cause	Corrective/preventive action
Cooling coil selection for low delta T than design	Always ensure that the cooling coil is selected for the same design delta T temperature before manufacturing the units and installing it at the site

10.3.4 Oversized Airside Equipment

Oversized fan coil units and air handling units receive more chilled water flow rate than the anticipated cooling load in the room it is serving, which leads to incomplete heat transfer between the chilled water and the room air that results in returning chilled water on a lower temperature than design. Corrective/Preventative action for this cause is mentioned in table 10.4:

Table 10.4 Oversized airside equipment.

Cause	Corrective/preventive action
Oversized airside equipment	Always ensure that airside equipment is selected according to cooling load demand

10.3.5 Chilled Water Pumps Selected with Higher Pump Head than Actual

If chilled water pumps are selected with higher pump head, the pumps will tend to pump more chilled water flow rate to the building during high load demand; the flow rate, in this case, will be more than what is required and will lead to return the water with lower temperatures than design return temperature. Corrective/Preventative action for this cause is mentioned in table 10.5:

Table 10.5 Chilled water pumps selected with higher pump head that actual.

Cause	Corrective/preventive action
Chilled water pumps selected with higher pump head than actual	Pumps should be selected on the right head, and index location should be determined in the chilled water circuit to ensure the right amount of chilled water flow rate is supplied to the building at all times

10.3.6 Use of Pressure-Dependent Control Valves

Pressure-dependent control valves are usually selected based on system pressure, which leads, in most cases, to valves being oversized. Oversizing allows more chilled water flow rate to circulate through the airside equipment according to system pressure, which results in lower return water temperature. Corrective/Preventative action for this cause is mentioned in table 10.6:

Table 10.6 Use of pressure-dependent control valves.

Cause	Corrective/preventive action
Use of pressure-dependent control valve	Use of pressure-independent control valves provides a steady flow for the airside equipment regardless of pressure fluctuation in the system, which always ensures that minimum flow is supplied to the airside equipment, and optimum delta T is achieved

10.4 Causes Can Be Avoided During Operation and Maintenance

Low delta T causes discussed in this section are a result of poor operation and maintenance. Those causes may be the root of the problem or may be contributed to it. Therefore, the maintenance personnel should conduct a detailed analysis to ensure that the root causes are known, and the associated corrective and preventive actions are put in place to ensure the issue will not happen again.

Preventive actions should be then implemented on a regular basis as maintenance inspections and tasks; those tasks should be defined in the planned maintenance job plans proactively on a predefined frequency. Maintenance frequency should be determined by the process mentioned in Chapter 6 in a way that prevents the problem from happening again.

The cost effectiveness of the process should be calculated based on all costs associated with the maintenance job plan compared to the energy saved and the waived low delta T charges imposed by cooling provider (if any).

10.4.1 Chilled Water Control Valve Left in Open Position

In some cases, it was found that the two-way chilled water control valve is left open position manually or stuck on open position, which prevents the control valve from doing its function in allowing the required chilled water flow rate from passing through the airside equipment, which causes overflowing of chilled water flow at part load conditions. In this case, the heat transfer between the air inside the room and the chilled water flowing inside the cooling coil is not complete which leads to low return chilled water temperature. Corrective/Preventative action for this cause is mentioned in table 10.7:

Table 10.7 Chilled water control valve left in open position.

Cause	Corrective/preventive action	Inspection frequency
Chilled water control valve left on open position	Check chilled water control valve position and ensure it is functioning on auto position and modulating chilled water flow according to demand Ensure control valve is not stuck on open position	Monthly

10.4.2 Chilled Water Control Valve is Not Responding to Space Temperature

In some cases, the two-way control valve does not modulate (close/open) by space temperature. If the set point temperature inside the room is 24.0 °C and the actual room temperature is also 24.0 °C, then the control valve actuator should close the two-way and prevent chilled water from flowing through the cooling coil.

If the control valve remained open, the chilled water flow would circulate inside the cooling coil with minimal heat transfer with the room air; the heat transfer will not be sufficient to warm the return chilled water temperature. Hence, water will return to the cooling plant lower than required and contributes to low delta T. Corrective/Preventative action for this cause is mentioned in table 10.8:

Table 10.8 Chilled water control valve is not responding to space temperature.

Cause	Corrective/preventive action	Inspection frequency
Chilled water control valve is not responding to space temperature	Check chilled water control valve position and ensure it is functioning on auto position and modulating chilled water flow according to demand Ensure control valve is not stuck on open position	Monthly

10.4.3 Dirty/Clogged Cooling Coil or Air Filter

If the cooling coil or the air filters of the airside equipment are clogged, the room air will not efficiently pass through the cooling coil. And the amount of the air that will pass through the coil will not be sufficient to complete the heat transfer between the chilled water passing inside the cooling coil, which will lead to low return chilled water temperature.

At the same time, since the airflow rate is lower than the commissioned airflow rate, the space set point temperature will not be achieved, which will maintain the control valve on open position trying to achieve the required room temperature, which will also allow the chilled water to pass through the cooling coil without sufficient heat transfer, and the return chilled water temperature remains lower than design.

Corrective/Preventative action for this cause is mentioned in table 10.9:

Table 10.9 Dirty/clogged cooling coil or air filter.

Cause	Corrective/preventive action	Inspection frequency
Dirty/clogged cooling coil or air filter	Check and clean cooling coil and air filters on a regular basis	Cleaning frequency differs between circulating airside equipment and fresh air units Fresh air units require a higher frequency of cleaning which could be bi-weekly or monthly

10.4.4 Opened Bypass Lines on Airside Equipment

During testing and commissioning stage, the bypass line is usually provided for flushing chilled water pipe network, which should be closed after flushing is done. In some cases, those bypass lines are left open which results in flowing the supply chilled water directly to the return lines with the same supply chilled water temperature. This is resulting in a high impact on return chilled water temperature and results in low delta T. Corrective/Preventative action for this cause is mentioned in table 10.10:

Table 10.10 Opened bypass lines on airside equipment.

Cause	Corrective/preventive action	Inspection frequency
Opened bypass lines on air side equipment	Check and ensure bypass lines are closed after flushing	Annual, or whenever system flushing takes place

11

Energy Centered Maintenance in Data Centers

11.1 ECM Terminology and Characteristics

Data centers' operations are usually 24 × 7 × 365 days a year. They need troubleshooting and monitoring all the time by both operations and maintenance. Keeping the servers running under the right environment without getting excessive heat or excessive water is the primary goal. They are declared by the organization as a mission critical facility. The data center can be in a facility by itself or in a facility containing other functions or organizations. It can be gigantic (multi-rooms) or located in one small room.

Data centers have HVAC, chillers, cooling towers, computer room air conditioning (CRACs), backup power (uninterruptable power supply – UPS), boilers, generators, lights, servers, and other equipment.

Power to the facility remains flat throughout the year (no seasonal changes as seen in other facilities) and costs around or more than $50,000 a month depending on size and equipment inside. Data centers use around 2% of energy consumed in the U.S. yearly.

The key definitions that apply to a data center are:

CRAC – computer room air condition.

PDU – power distribution units: distributes power to servers, performs power filtering and load balancing, and provides remote monitoring and control.

Plenum – concealed area that enables hot air flow back to CRAC.

PRO DC – a software program developed by DOE to identify energy savings in a data center.

PUE – Power Use Effectiveness = Total Facility Power/IT Equipment and Operations Use in kW.

Rack – container that holds the servers.

Raised floor – enables the cool air or cold water cables to cool the servers.
Server – a computer that computes or provides service to a network or Internet.
UPS – uninterruptable power supply: provides backup power in case of an interruption to main power supply.

The data center uses a lot of energy. ECM is a natural tool to use to decrease energy consumption while ensuring that data center operations is productive and safe from the dangerous items that can quickly bring a data center to a halt.

If your data center is cold, then these solutions will make you gold.

If the data center is cold, then ECM is needed immediately to reduce energy consumption from 30% to 40%. Your approach is six actions.

Action 1. Place the servers in a hot/cold configuration with only cold air on one side and hot air coming from the server back on the other side.

Action 2. Check the temperatures in the server room. It is probably around 66°F. The ASHRAE Standard allows significantly higher temperatures now than in the past. By raising the temperatures, less air conditioning is needed, less work on the cooling towers, and you may be able to retire one or more CRACs since they may not be needed anymore. Many data centers are too cold and dry. In 2008, ASHRAE adjusted its guidance around supply air temperatures to 64.4°F at the lowest extreme and 80.6°F at the high end (up from its previous recommended high of 77°F). Many data centers have been setting temperatures much lower, as low as 55°F. By raising the temperature by even 1°, data centers can achieve energy savings. For humidity, ASHRAE has also loosened its guidelines, increasing the high end to 60% relative humidity (up from 55% in 2004).

Action 3. Separate the cold air from mixing with the hot air on the way back to the CRAC. You may need a contractor to plan and have this built and put in place. Panels are used including panel doors to seal off the cold aisles. This project will have less than two years' payback. Be sure to include the fire marshal and the maintenance department or contractor from the beginning of the planning and execution.

Action 4. Using a wattmeter, find out for each server the watts being used to power the server and compare to the nameplate what should be utilized. The old servers could easily be "energy hogs" and, if so, should be replaced.

Action 5. Measure the utilization of each server. If the utilization is low, then have the programmers increase the use. This action will provide more server efficiency, thus reducing some servers for being used and lowering the energy consumed. There will need to be some servers with lower utilization to be used as a backup to the main ones.

Action 6. Calculate either with actual data or estimated data, the power usage effectiveness (PUE). Power Usage Effectiveness = Total Facility Power (kW)/IT Equipment Power (kW). A PUE of 1.6 to 1.3 is good, with 1 being the goal. The lower the number, the larger percentage of the total power furnished to the facility is used by the data center operations. (80% of the power furnished used by the IT equipment would be excellent. This is the DCIE indicator whose reciprocal is the PUE. 1/0.8 = 1.25.)

*Facebook's new data center in Forest City, NC, USA has a PUE between 1.06 and 1.08.

*A high PUE should drive improvement in your data center.

Taking these six actions will lead to gold or at least a lot of energy and money saved.

12

Measures of Equipment and Maintenance Efficiency and Effectiveness

12.1 Lead (Key Performance Indicators) and Lag (Key Result Indicators)

Lagging indicators are normally "output" or "outcome" oriented, easy to measure but difficult to improve while leading indicators are normally "input" oriented, challenging to measure and easy to influence. Lead Indicators are measures that **"drive"** the **performance** of lag ones. Normally, lead indicators measure processes and activities (figures 12.1, 12.2). Lag indicators are outcomes of these processes. The lead indicators are often called key performance indicators (KPIs).

KRIs tell management and stakeholders if the organization is operating well. They do not provide any information or insight into what activities or actions were successful or not. They do determine whether the organization's end goals were met or not.

While KRIs cannot help you achieve or improve upon the organization's goals, KPIs focus on the actions that lead to the results. KPIs provide the information that is important in creating strategies and aligning goals.

Key result indicators (KRIs) measure *the results of goals and business strategies (figure 12.3)*. Key performance indicators (KPIs) measure actions that are important and influence or impact the **KRIs**. Most organizations need a mixture of *KPIs* and *KRIs*. Whether an organization meets their corporate goals mostly depends on whether there are helpful KPIs to influence the KRI outcomes.

Figure 12.1 Introduction to lead and lagging indicators.

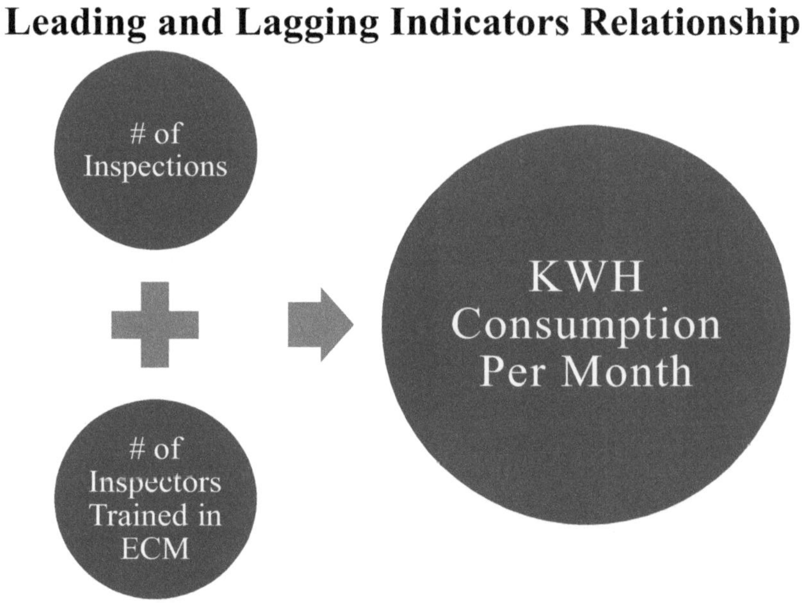

Figure 12.2 Lead and lag indicators relationships.

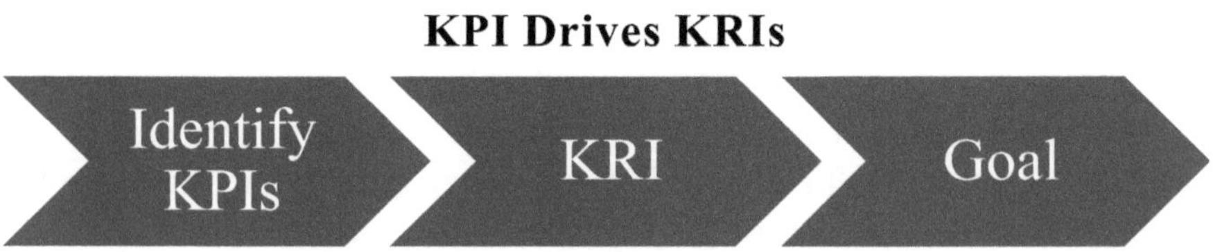

Figure 12.3 KPI (lead) drives KRIs (lag).

12.2 Maintenance Group Indicators

The maintenance group or department should have a combination of KPIs and KRIs. The KRIs are results and normally are traditional indicators. The KPIs drive the KPIs and are drivers and as such are very actionable. Once they are completed, they are no longer actionable and should be discontinued. For example, percent people committed. Once everyone is committed, we no longer need to measure and display this KPI. We now need to find another, if possible, that will drive the particular KRI.

The maintenance department or group should have measures they are in control of and are actionable only by having a KPI to influence the KRIs favorably. It is so easy to select performance indicators that provide information but are not actionable that can help improve the maintenance operations or equipment reliability. Some traditional maintenance indicators such as mean time between failure (MTBF) are impacted by others such as affected by the ordering and receiving of the original equipment, the manufacture's quality of production, the equipment design, maintenance tasks performed or not performed, and quality of the parts used and even workmanship.

The planning group should already have some performance indicators before implementing energy centered maintenance (ECM). Some of these could be impacted by ECM and be impossible to distinguish if the preventative maintenance or predictive maintenance achieved the results or ECM. Therefore, the ECM performance indicators selected should be controlled in the ECM process and be actionable to the ECM goals and objectives.

Setting goals are important for any maintenance group to improve their efficiency, productivity, and effectiveness. The ECM technicians or mechanics will have to abide by the rules and be a part of the operations helping achieve these goals. What are these goals? Of course, these goals will vary by organization, but several will probably be included by most organizations. They are:

Goal 1: Up time greater than 90%.
Goal 2: Percent emergency work less or equal to 10%.
Goal 3: Training days per technician or mechanic is 6–9 days a year.
Goal 4: Schedule achievement is greater than 90%.
Goal 5: Technicians or mechanics to a maintenance planner 25 to 1.
Goal 6: Percent overtime less than 8%.
Goal 7: Maintenance cost to estimated equipment replacement cost is 2.5% or less.

The technicians or mechanics performing ECM tasks will have some impact, although not the primary purpose, on Goal 1. They will adhere to Goal 3, 6, and 7, help achieve Goal 4, and be included in Goal 5. At least four performance measures jump out from these goals. Energy centered maintenance indicator #1 is schedule achievement. The indicator target is 90% or higher. The second indicator, ECM indicator #2, is less than 8% overtime. Since the ECM tasks are not fixing breakdowns, overtime should rarely be used. There will be occasions where overtime may be necessary to inspect a machine or repair it on off-peak hours. However, the target should be 1% or less, considerably lower than 8% for other maintenance. ECM performance indicator #3 is ECM cost divided by the estimated energy savings from doing the inspections, maintenance, repair, or replacement. This ladder indicator is an ECM payback indicator with a target of 3 years or less. ECM indicator #4 is % of technicians or mechanics trained in ECM. The target is 100% in the first six months.

The work done by maintenance needs to support the business aims and operating strategy. The ideal way to show that is to have maintenance performance linked to the reasons your company is in business. This is harder for most other maintenance types to do than for ECM in that every organization needs the energy to perform its mission providing its products or services. The energy costs can be substantial, sometimes up to 30% of the total operational costs. Therefore, the energy cost reduction (KWH Reduced × Cents Per kWh = Cost Savings) is a welcomed initiative by any company, university, or business. Performance indicators #5 and #6 are energy consumption reduced and energy costs savings, respectively. Along with these, the key result indicator, electricity or energy intensity, becomes our ECM key results indicator #7. The target is 5% reduction during the first year. It is defined for electricity as KWH consumed/total gross square feet of facilities maintained.

Effective maintenance is defined as doing what is right. Training of technicians or mechanics and having detailed precise maintenance instructions coupled with work order planning and scheduling will favorably impact the maintenance effectiveness. Doing efficient maintenance is doing the right work required that improves reliability or accomplishes its purpose with the least resources and with productivity. Successful maintenance is not fixing things but not having to fix them. Therefore, not having to do the maintenance would be an excellent indicator showing the maintenance effectiveness. This is a useful performance indicator but tough to measure. It is important that when we select a group of maintenance key performance indicators, we pick the ones that let us improve equipment reliability, maintenance performance, workforce productivity, and goals achievement, and not just tell us the problems experienced in our businesses or organizations.

ECM can contribute to equipment reliability improvement, even though that is not its primary objective or purpose. Some organizations may want a key result indicator to measure if equipment reliability is being improved or not in the long term. For example, some equipment inspected in ECM that were targeted for possible energy excessive use could measure mean time between failure for these. The industry mean time between failure could be obtained (ECM indicator #8 MTBF). The target is more than three years. Then, measuring when they have an actual failure could show whether equipment reliability has been improved or not. For example, the equipment to be measured could have a mean time between failure as shown in table 12.1.

Table 12.1 Targeted equipment with present MTBF.

Equipment	MTBF (To be verified with OEM)
Pumps	3 Years
Compressors	4 Years
Air handling units	3 Years
Boilers	2.5 Years
cooling towers	2.5 Years
Elevators	2 Years
Water supply system	3 Years

KRI ECM indicator #9 is percent improvement in five years (could be any year after 4) and can be calculated for individual equipment or as an aggravation. The target is 10% increase. This is an optional key result indicator.

For measuring maintenance, the areas that often are measured are:

1. The efficiency of the maintenance group.
2. The effectiveness of the maintenance group.
3. The productivity of the maintenance group.
4. The performance of the maintenance group.
5. Progress against maintenance goals or adherence to maintenance policies.
6. Plant reliability goals.
7. Equipment availability along with down time.
8. Schedule achievement.
9. Business value of maintenance.
10. The quality of maintenance.
11. Overall equipment efficiency.

12.3 Overall Equipment Efficiency (OEE)

KRI #10 is OEE = Availability × Performance × Quality

http://www.oee.com/calculating-oee.html

Let us say the target is 87%. These three areas are usually measured separately and then multiplied to give the overall equipment efficiency. Therefore, individual targets can be established for each of the three factors and when they are multiplied to get OEE would equal 87% (providing the target is achieved).

The reasons the equipment's OEEs are low is due to extensive setup and cleaning periods, production downtime, inefficient processes, and poor planning.

These problem areas can be removed by improving the transparency of machine utilization times, collecting and analyzing potential sources of loss, and always providing information on production processes in real time.

KRI #11 is Machine Efficiency:

Machine efficiency is:

Efficiency = Output Value/Input Value

For example, if a machine needs 10 kW to run and produces 8 kW, its power efficiency is 8/10 = 0.8 or 80%.

Efficiency is always between 0 and 1 (or 0 and 100 if expressed as a percentage).

ECM KRI #12 is OEE availability:

Availability includes all events or happenings that stop planned production sufficient enough where it makes sense to track a reason for being down (several minutes).

Availability is defined as the ratio of run time to planned production time where planned production time is the total time that equipment is expected to produce.

Availability takes into account down time loss and is calculated as:

Availability = Operating Time/Planned Production Time.

Down time can be a result of lack of maintenance, ECM, or preventative or predictive.

ECM KRI 13 is OEE Performance:

The performance takes into account speed loss. Its formula is:

Performance = Ideal Cycle Time/(Operating Time/Total Pieces).

Ideal cycle time is the minimum cycle time that your process can be expected to achieve in optimal circumstances. It is sometimes called *design cycle time*, *theoretical cycle time*, or *nameplate capacity*.

Since run rate is the reciprocal of cycle time, *performance* can also be determined as:

Performance = (Total Pieces/Operating Time)/Ideal Run Rate.

ECM KRI # 14 is Quality:

Quality takes into account quality loss and is calculated as:

Quality = Good Pieces/Total Pieces.

OEE scores provide very useful information. It is an accurate picture of how effectively your manufacturing process is running. In addition, it enables improvements to be tracked over time.

The OEE score by itself does not provide any insights as to the underlying causes of lost productivity. This is the role of availability, performance, and quality.

In the preferred calculation, you get the best of both worlds. A single number that captures how well you are doing (OEE) and three numbers that capture the fundamental nature of your losses (availability, performance, and quality).

Here is an interesting example described in table 12.2. Look at the following OEE data for two sequential weeks.

Table 12.2 Overall equipment efficiency for two sequential weeks.

OEE factor	**Week 1**	**Week 2**
OEE	77.1%	78.1%
Availability	88.0%	88.1.%
Performance	89.4%	89.81%
Quality	98.0%	97.92%
Item	**Data**	
Shift length	8 hrs (480 minutes)	
Breaks	(2) 15 minutes and (1) 30 minutes	
Down time	43 minutes	
Ideal cycle time	1.0 seconds	
Total count	20,100 do dads	
Reject count	418 do dads	

- Planned production time:

The OEE calculation begins with planned production time. So, first, exclude any shift time where there is no intention of running production (typically breaks).

Formula: Shift Length – Breaks.

Example: 480 minutes – 60 minutes = 420 minutes.

- Run time:

The next step is to calculate the amount of time that production was actually running (was not stopped). Remember that stop time should include both unplanned stops (e.g., Breakdowns) and planned stops (e.g., Changeovers). Both provide opportunities for improvement.

Formula: Planned Production Time – Stop Time.

Example: 420 minutes – 43 minutes = 377 minutes.

- Good count:

If you do not directly track good count, it also needs to be calculated.

Formula: Total Count – Reject Count.

Example: 20,100 do dads – 418 do dads = 19,682 do dads.

- Availability:

Availability accounts for when process is not running (both unplanned stops and planned stops).

Formula: Run Time/Planned Production Time.

Example: 373 minutes/420 minutes = 0.8881 (88.81%).

- Performance:

Performance accounts for when the process is running slower than its theoretical top speed (both small stops and slow cycles).

Formula: (Ideal Cycle Time × Total Count)/Run Time.

Example: (1.0 seconds × 20,100 do dads)/(373 minutes × 60 seconds) = 20,100/22,380 = 0.8981 (89.81%).

- Quality:

Quality accounts for manufactured parts that do not meet quality standards.

Formula: Good Count/Total Count.

Example: 19,682/20,100 do dads = 0.9792 (97.92%).

- OEE:

Finally, OEE is calculated by multiplying: Availability × Performance × Quality.

Example: 0.8881 × 0.8981 × 0.9792 = 0.78.1 (78.1%).

OEE can also be calculated using the simple calculation.

Formula: (Good Count × Ideal Cycle Time)/Planned Production Time.

Example: (19,682 widgets × 1.0 seconds)/(420 minutes × 60 seconds) = 19,682/25,200 = 0.781 (78.1%).

The result is the same in both cases. The OEE for this shift is 78.1%.

12.4 ECM Inspection

The schedule achievement was measured in KRI #1. We also need to know the percentage of equipment's performance measured on a certain decided time (KPI #15) and when the equipment is using excessive energy, how many, or percent were fixed in 1 day or less (KPI #16).

12.5 Indicator Checked

We know the reasons for indicators are developed. We can use a table or matrix and check to see if we adequately covered the important areas and see if we have any serious gaps in our measurement coverage. Table 12.3 provides a sample indicator check.

Table 12.3 Indicators check.

Reasons for indicators	KPI (lead)	KRI (lag)	Coverage adequate Yes or No?	Possible new indicator or one needed
1. The efficiency of the maintenance group	4. % Trained in ECM; 16. % equipment using excessive energy fixed in 1 day or less	11. Machine efficiency	Yes, for ECM, No for other maintenance	% Tools requested and filled (KPI)
2. The effectiveness	15. % Equipment inspected on time; 16. % equipment using excessive energy fixed in 1 day or less	5. Energy consumption; 6. energy cost; 7. energy intensity; 8. MTBF; 9. % improvement in MTBF; 12. availability and down time	Yes	
3. The productivity			Not needed for ECM	
4. The performance	15. % Equipment inspected on time; 16. % equipment using excessive energy fixed in 1 day or less	10. OEE	Yes	
5. Goal achievement progress or adherence to policy	15. % equipment inspected on time; 16. % equipment using excessive energy fixed in 1 day or less	2. Over time	Yes	Need a policy and procedure
6. Plant reliability	15. % Equipment inspected on time, % Equipment using excessive energy fixed in 1 day or less	8. MTBF; 9. % improvement in MTBF	Yes for ECM, No for Other maintenance	Not ECM's main purpose

Table 12.3 Continued

Reasons for indicators	KPI (lead)	KRI (lag)	Coverage adequate Yes or No?	Possible new indicator or one needed
7. Equipment availability	15. % Equipment inspected on time; 16. % equipment using excessive energy fixed in 1 day or less	12. Down time and availability	Yes	
8. Schedule achievement	15. % Equipment inspected on time; 16. % equipment using excessive energy fixed in 1 day or less	1. Schedule achievement; 11. machine efficiency 13 OEE performance	Yes	
9. Business value of maintenance		3. Payback; 5. energy consumption; 6. energy cost; 7. energy intensity	Yes	
10. Quality of maintenance		14. Quality	Yes	
11. Overall equipment efficiency	15. % Equipment inspected on time, % equipment using excessive energy fixed in 1 day or less	10. OEE	Yes	

It is easy to get a flood of indicators. Be sure only to select those that are meaningful and develop a data collection sheet for each.

Data collection sheet:

1. Indicator Title:
2. Data Source:
3. Frequency:
4. Formula:
5. Graph Type:
6. Description of Data:
7. Person Responsible:

Most of the areas are self-explanatory except the Description of Data. For example, if data is included for weekends or not included, people would know by this description. Each indicator should have a data collection sheet.

12.6 Target Setting

Setting a target is not about magically pulling a figure out of the air. To set a target, you must first know where you are, what you want to achieve, and then be able to determine realistically, but stretch estimates that represent challenging amounts of improvement needed to attain the target.

How do we define a target? A target and stretch target can be defined as follows.

Targets: The desired level of performance you want to achieve on your indicator after a period that shows success from the countermeasures you have implemented along the way.

Stretch target: A challenging but an attainable target that can be reached with reasonable but slightly accelerated effort.

The steps in setting a target are:

1. Define what level or present status.
2. Determine "How Much" or What You Want to Achieve and "By When."
3. Establish the timeframes you need the target to be attained.

In setting the targets, it is good to set a SMART one by using and adhering to the SMART characteristics:

- **S**pecific: What you plan to achieve is understood.
- **M**easurable: There is an indicator or measure that shows whether you have achieved it or not.
- **A**chievable: With the resources available.
- **R**ealistic: Stretch, but possible.
- **T**imeframe: From and to are specified.

Targets drive performance. Therefore, set good ones but make them achievable. Targets and indicators go together. With every good indicator, there is always a good target.

13

Energy Savings Verification

13.1 Baseline

The typical way energy savings is achieved by implementing certain measures, or executing an energy saving project that helps achieving the targeted KWH reduction. Those reductions are then should be verified as savings when compared to the baseline energy consumption values

Having a baseline is critical. It gives you something to compare the current to see if any improvement has been made or not.

Energy baseline is the basis for an initial energy assessment of the building; it summarizes the building's current energy performance and helps the energy management team to draw basis of where the complete energy management processes should start. It will also assist the energy management team to understand how energy expenditure contributes to operating cost.

To develop a baseline, obtain three years of electricity bills and plot the monthly consumption on a line graph. Energy baseline can be developed and evaluated by energy management team based on energy data history and real-time metering. An energy baseline is an approach in which all energy inputs to a facility are accounted for; it is a preliminary energy assessment for the current energy behavior of the building before conducting detailed energy analysis. Energy baseline is the analysis of the energy use and consumption based on measurement and other data obtained from metering and energy bills.

The information extracted from the preliminary energy baseline assessment will allow the energy management team to benchmark the building energy performance with other buildings, establishing energy performance indicators and setting energy management targets.

Energy baseline will help the energy management team to:

- Identify the current energy consumption of the facility.
- Establish internal energy consumption benchmark.

- Benchmark energy consumption with other similar facilities.
- Identify of all internal and external factors affecting the energy consumption of the facility.
- Identify of significant energy consumption equipment.
- Identify of energy flows in the facility.
- Estimate future energy use and consumption.
- Identify, prioritize, and highlight opportunities for improvement.

Energy baseline puts the basis to ask how we can reduce electricity consumption. We can reduce either the connected load or the operating hours. Finding motors running 24 hrs a day when they were only required to run 7 hrs will reduce the operating load for motors. In universities and colleges, during holidays and times when rooms are not occupied, the equipment in the room, the air conditioning, or heating and ventilation can be turned off, saving energy consumption. Check each item's necessary operating time and, if smaller, reduce the operator time. Changing operation hours for some equipment from peak hours to off-peak can save considerable energy costs. When replacing any equipment, it may be possible to reduce the connected load by purchasing a more energy-efficient one such as Energy Star and replacing the item when it uses excessive energy or about to break down.

13.2 Example of Energy Baseline

To develop a baseline, obtain three years of electricity bills and plot the monthly consumption on a line graph. For example, the data in table 13.1 represent the monthly energy consumption for years 2013, 2014, and & 2015, which are used to plot the line graph represented in figure 13.1.

Table 13.1. kWh by month and year.

	Year/month			
	2013	**2014**	**2015**	**3 years monthly average**
Jan	42,090	43,000	39,800	41,630
Feb	44,090	43,987	40,378	42,818
Mar	39,800	40,122	40,000	39,974
Apr	38,678	38,980	36,789	38,149
May	38,900	39,100	37,500	38,500
June	41,267	40,900	39,870	40,679
July	43,300	44,600	41,380	43,093
Aug	44,500	44,190	44,200	44,297
Sept	46,780	46,000	44,560	45,780
Oct	45,189	44,236	42,900	44,108
Nov	46,780	46,700	44,360	45,947
Dec	47,990	46,789	44,350	46,376

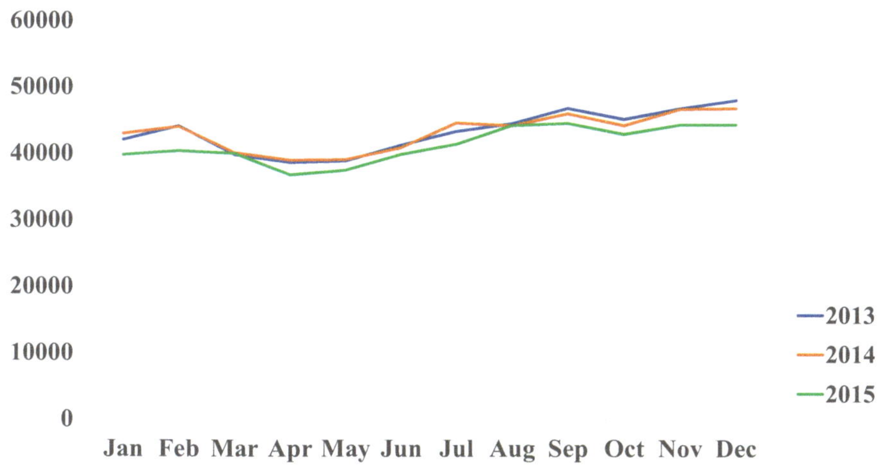

Figure 13.1 kWh monthly consumption by year.

Notice the seasonality of the yearly graph (figure 13.1), s and the total yearly consumption in table 13.2. Years 2013 and 2014 have almost same trends. The year 2015 shows some reduction of kWh consumption. In late 2014, some energy projects were implemented, and the resulting reduction is happening. The percent improvement is 518,595 – 496,087 = 22,508/518,595 =.043 = 4.3%.

Table 13.2. Year totals.

2013	2014	2015
519,364	518,595	496,087

Therefore, either 2013 or 2014 would be a better baseline. However, let us see how the yearly monthly averages compare to 2014.

The year 2014 and the monthly averages trend are almost identical. Therefore, the year 2014 is selected as the baseline that all possible improvements in the future will be compared with for verification.

How can we reduce electricity consumption? We can reduce either the connected load or the operating hours. Finding motors running 24 hrs a day when they were only required to run 7 hrs will reduce the operating load for motors. In universities and colleges, during holidays and times when rooms are not occupied, the equipment in the room, the air conditioning, or heating and ventilation can be turned off, saving energy consumption. Check each item's necessary operating time and, if smaller, reduce the operator time. Changing operation hours for some equipment from peak hours to off-peak can save considerable energy costs. When replacing

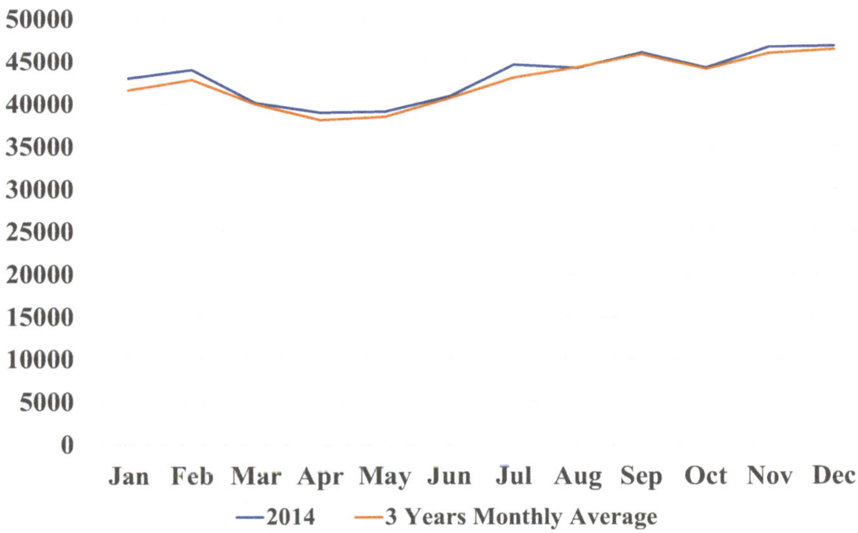

Figure 13.2 Monthly Energy Consumption Trend (Years 2013, 2014)

any equipment, it may be possible to reduce the connected load by purchasing a more energy-efficient one such as energy star and replacing the item when it uses excessive energy or is about to break down.

13.3 Energy Baseline, Energy Targets, and Energy Performance Indicators

Developing energy baseline alone does not enable the energy management team to determine if energy is being used efficiently on a continuing basis or not. Developing energy baseline should be part of the operating culture of the facility and should be developed on a regular basis to ensure that energy is consumed in an efficient way on a continuing basis. Therefore, energy baseline should be transformed to monthly energy targets and energy performance indicators (EnPI's) which should be developed for the building by the energy management team.

Setting energy targets is important because it provides a direction of what needs to be achieved within a certain time frame. Therefore, after energy benchmarking is set and energy performance indicators are defined and analyzed, the energy management team shall be able to set energy targets to improve performance. Energy targets must be measurable and achievable depending on information from benchmarking.

Energy targets shall identify target levels including what savings are included, how savings are to be evaluated, and how specific energy performance indicators and baselines are to be used. The moment the energy saving targets are set, the effect of energy management process shall start by conducting the energy audit. Further, where energy targets are found to be achievable via systems retrofit, the energy management team shall develop a business case feasibility study comparing the investment cost with the return-on-investment value over a specified period to confirm its feasibility.

The ability to measure the real performance is the foundation for improvement and optimization.

Energy management key performance indicators are used to track the percentage of organization's goals for energy and carbon reduction.

Three performance indicators are critical for managing energy consumption: energy utilization index (EUI) – table 13.3, energy cost index (ECI) – table 13.4, and energy productivity index (EPI) – table 13.5. These indicators provide a measure of which the degree of performance can be related to a given target

Energy Utilization Index (EUI): Ratio of the energy consumed in a building over a given period (year) per gross square meter of conditioned space.

Table 13.3 Energy utilization index (EUI).

Value driver	Unit	Measurement	Energy reduction target
Energy utilization index (EUI)	kWh/m²/year	The EUI is the measure of the total kWh used per year to the total number of square meters of conditioned space	Site-specific reduction (%) compared to previous year EUI (downward trend)

Energy Cost Index (ECI): Measure of the total energy cost of a building over a period of time (year) per gross square meter of conditioned space.

Table 13.4 Energy cost index (ECI).

Value driver	Unit	Measurement	Energy reduction target
Energy cost index (ECI)	$/m²/year	The ECI adds up all costs of energy and divides the result by total square meters of conditioned space	Site-specific reduction (%) compared to previous year ECI (downward trend)

Energy Productivity Index (EPI): Measure of the total energy consumed in a building over a period (year) per occupant or visitor, expressed in kWh/capita/year.

Table 13.5 Energy productivity index (EPI).

Value driver	Unit	Measurement	Energy reduction target
Energy productivity index (EPI)	kWh/capita/year or $/capita/year	The EPI adds up all energy consumed (or billed) by the facility divided by the number of visitors or occupants	Site-specific reduction (%) compared to previous year EPI (downward trend)

13.4 Energy Centered Maintenance and Energy Performance Indicators

Energy performance indicators in energy centered maintenance model can be measured as part of the building's EUI, ECI, and EPI. However, those indicators provide information about how the entire facility is performing from an energy perspective and how each energy critical equipment is performing.

Therefore, new energy performance indicators need to be developed for each equipment by itself to measure its performance before and after the energy maintenance tasks took place and were implemented. Those indicators differ according to equipment type and the type of energy it consumes (for example, measuring motor power consumption in an air handling unit before and after conducting the maintenance tasks, and measuring the effectiveness of water-to-water heat exchanger).

Building's energy performance indicators focus on energy consumption of the building; information about total energy consumption is extracted from energy bills or energy meters. On the other side, energy performance of equipment should be measured by the maintenance personnel while conducting energy centered maintenance tasks. Information about heat exchangers' effectiveness need to be measured and calculated manually, and information about air handling unit's energy consumption can be extracted from a meter that is measuring the AHU motor energy consumption.

Energy performance indicators related to equipment performance involve collecting and analyzing actual data about the operational parameters of the equipment such as equipment's efficiency, equipment's effectiveness, power consumption, etc.

The following equipment's performance indicators (through tables 13.6 to 13.21) can be used to measure the equipment's performance before and after energy centered maintenance tasks are conducted to measure the actual energy savings achieved.

Table 13.6 Performance indicator for air handling units.

Measurement	Performance indicator	Target
Supply/return airflow rate (m^3/hr)	% of actual flow rate/T&C flow rate	±5% of testing and commissioning value
Full load motor's electricity consumption (kWh)	% of actual energy consumption/ T&C energy consumption	±5% of testing and commissioning value
Full load cooling coil pressure drop (PSI)	% of actual pressure drop/T&C pressure drop	±5% of testing and commissioning value
Full load chilled water temperature difference (°C)	% of actual delta T/design delta T	± 2% of design and commissioned temperature difference (°C)

Table 13.7 Performance indicator for fan coil units.

Measurement	Performance indicator	Target
Supply/return airflow rate (m^3/hr)	% of actual flow rate/T&C flow rate	±5% of testing and commissioning value
Full load motor's electricity consumption (kWh)	% of actual energy consumption/ T&C energy consumption	±5% of testing and commissioning value
Full load cooling coil pressure drop (PSI)	% of actual pressure drop/T&C pressure drop	±5% of testing and commissioning value
Full load chilled water temperature difference (°C)	% of actual delta T/design delta T	±2% of design and commissioned temperature difference (°C)

Table 13.8 Performance indicator for energy recovery units.

Measurement	Performance indicator	Target
Supply/return airflow rate (m^3/hr)	% of actual flow rate/T&C flow rate	±5% of testing and commissioning value
Full load motor's electricity consumption (kWh)	% of actual energy consumption/ T&C energy consumption	±5% of testing and commissioning value
Full load cooling coil pressure drop (PSI)	% of actual pressure drop/T&C pressure drop	±5% of testing and commissioning value

Table 13.9 Performance indicator for boilers.

Measurement	Performance indicator	Target
Fuel combustion	% of actual fuel consumption rate/T&C fuel consumption rate	±5% of testing and commissioning value

Table 13.10 Performance indicator for pumps.

Measurement	Performance indicator	Target
Full load water flow rate (gpm, lps)	% of actual flow rate/T&C Flow rate	100%–110% of testing and commissioning value
Full load motor's electricity consumption (kWh)	% of actual energy consumption/ T&C energy consumption	±5% of testing and commissioning value

Table 13.11 Performance indicator for close control units.

Measurement	Performance indicator	Target
Supply/return airflow rate (m^3/hr)	% of actual flow rate/T&C flow rate	±5% of testing and commissioning value
Full load cooling coil pressure drop (PSI)	% of actual pressure drop/ T&C pressure drop	±5% of testing and commissioning value
Full load fan motor's electricity consumption (kWh)	% of actual energy consumption/T&C energy consumption	±5% of testing and commissioning value
Full load compressor's electricity consumption (kWh)	% of actual energy consumption/T&C energy consumption	±5% of testing and commissioning value
Full load chilled water temperature difference (°C)	% of actual delta T/design delta T	±2% of design and commissioned temperature difference (°C)
DX unit energy efficiency ratio	EER = cooling capacity (Btu per hour)/power input (Watts)	Matching original EER as per manufacturer records

Table 13.12 Performance indicator for fans.

Measurement	Performance indicator	Target
Supply/return airflow rate (m^3/hr)	% of actual flow rate/T&C flow rate	±5% of testing and commissioning value
Full load motor's electricity consumption (kWh)	% of actual energy consumption/T&C energy consumption	±5% of testing and commissioning value

Table 13.13 Performance indicator for cooling towers.

Measurement	Performance indicator	Target
Supply/return airflow rate (m^3/hr)	% of actual flow rate/T&C flow rate	± 5% of testing and commissioning value
Full load motor's electricity consumption (kWh)	% of actual energy Consumption/T&C energy consumption	± 5% of testing and Commissioning value
Cooling tower range at full load (°C)	In–out water temperature	±5% of testing and commissioning value

Table 13.14 Performance indicator for air cooled chillers.

Measurement	Performance indicator	Target
Full load evaporator pressure drop (PSI)	% of actual pressure drop/T&C pressure drop	±5% of testing and commissioning value
Full load condenser fans motor's electricity consumption (kWh)	% of actual energy consumption/T&C energy consumption	±5% of testing and commissioning value

Measurement	Performance indicator	Target
Full load compressor's electricity consumption (kWh)	% of actual energy consumption/ T&C energy consumption	±5% of testing and commissioning value
Full load chilled water temperature difference (°C)	% of actual delta T/design delta T	±2% of design and commissioned temperature difference (°C)
Energy efficiency ratio	EER = cooling capacity (Btu per hour)/power input (Watts)	Matching original EER as per manufacturer records

Table 13.15 Performance indicator for plate heat exchangers.

Measurement	Performance indicator	Target
Full load water pressure drop (PSI)	% of actual pressure drop/T&C pressure drop	±5% of testing and commissioning value
Full load water temperature difference (T hot-in – T Cold-in) (°C)	Difference between the inlet temperature on the hot water side minus the inlet cold water temperature on the cold side	±2% of design and commissioned temperature difference (°C)

Table 13.16 Performance indicator for water cooled chillers.

Measurement	Performance indicator	Target
Full load evaporator pressure drop (PSI)	% of actual pressure drop/T&C pressure drop	±5% of testing and commissioning value
Full load condenser pressure drop (PSI)	% of actual pressure drop/T&C pressure drop	±5% of testing and commissioning value
Full load compressor's electricity consumption (kWh)	% of actual energy consumption/T&C energy consumption	±5% of testing and commissioning value
Full load chilled water temperature difference (°C)	% of actual delta T/design delta T	± 2% of design and commissioned temperature difference (°C)
Energy efficiency ratio	EER = cooling capacity (Btu per hour)/power input (Watts)	Matching original EER as per manufacturer records

Table 13.17 Performance indicator for water cooled chillers.

Measurement	Performance indicator	Target
Full load compressor's electricity consumption (kWh)	% of actual energy consumption/ T&C energy consumption	±5% of testing and commissioning value
Full load chilled water temperature difference (°C)	% of actual delta T/design delta T	±2% of design and commissioned temperature difference (°C)
Energy efficiency ratio	EER = cooling capacity (Btu per hour)/power input (Watts)	Matching original EER as per manufacturer records
Full load condenser fans motor's electricity consumption (kWh)	% of actual energy consumption/T&C energy consumption	±5% of testing and commissioning value

Table 13.18 Performance indicator for pressure reducing valve stations.

Measurement	Performance indicator	Target
Water pressure after PRV	% of actual water pressure/T&C water pressure record	±5% of testing and commissioning value

Table 13.19 Performance indicator for travelators, lifts, and escalators.

Measurement	Performance indicator	Target
Full load motor's electricity consumption (kWh)	% of actual energy consumption/ T&C energy consumption	±5% of testing and commissioning value

Table 13.20 Performance indicator for motor control centers.

Measurement	Performance indicator	Target
In–out voltage records	In–out voltage readings	100% matching testing and commissioning value in case no changes happened on electrical load connected to MCC

Table 13.21 Performance indicator for variable frequency drives.

Measurement	Performance indicator	Target
In/out current at constant frequency	In/out current at constant frequency readings	100% matching testing and commissioning value in case no changes happened on electrical load connected to MCC

13.5 Savings in Data Center Measures and Verification

If possible, a separate electricity meter for the data center will help measure the consumption and whether it is reduced or not through implementing energy reduction countermeasures. The main measure to identifying whether energy reductions occurred is the kWh per month consumption and the electricity intensity (kWh/Sq Ft. of the data center. The power usage effectiveness (PUE) is still a good measure of effectiveness, but these first two are the best for energy improvements verification.

Action 1. Place the servers in a hot/cold configuration with only cold air on one side and hot air coming from the server back on the other side.

Measures for Verification: kWh consumption per month and electricity intensity are the best verification measures.

Action 2. Check the temperatures in the server room. It is probably around 66°F. The ASHRAE Standard allows significantly higher temperatures now than in the past. By raising the temperatures, less air conditioning is needed, less work on the cooling towers, and you may be able to retire one or more CRACs since they may not be needed anymore. Multiple data centers are found to be too cold and dry. In 2008, ASHRAE issued a revision of its guidance around the temperature of supply air to 64.4°F being lowest acceptable and 80.6°F highest (up from its previous recommended high of 77°F). Many data centers have been setting temperatures much lower, as low as 55°F. By raising the temperature by even 1°, data centers can achieve energy savings. For humidity, ASHRAE has also loosened its guidelines, increasing the high end to 60% relative humidity (up from 55% in 2004).

Measures for Verification: kWh consumption per month and electricity intensity are the best verification measures.

Action 3. Separate the cold air from mixing with the hot air on the way back to the CRAC. You may need a contractor to plan and have this built and put in place. Panels are used including panel doors to seal off the cold aisles. This project will have less than two years' payback. Be sure to include the fire marshal and the maintenance department or contractor from the beginning of the planning and execution.

Measures for Verification: kWh consumption per month and electricity intensity are the best verification measures.

Action 4. Using a wattmeter, find out for each server the watts being used to power the server and compare to the nameplate what should be used. The old servers could easily be "energy hogs" and, if so, should be replaced.

Measures for Verification: Number of servers using excessive energy/total number servers inspected × 100 = percent servers inspected that are using excessive energy.

Action 5. Measure the utilization of each server. If the utilization is low, then have the programmers increase the use. This task will provide more server efficiency, thus reducing some servers for being used and lowering the energy consumed. There will be need of some servers with lower utilization to be used as a backup to the main ones.

Measures for Verification: Servers Utilization = Sum of $N1 + N2 + \ldots + Nt$/Total Number of Servers × 100.

Action 6. Calculate either with actual data or estimated data, the power usage effectiveness (PUE). Power Usage Effectiveness = Total Facility Power (kW)/IT Equipment Power (kW). A PUE of 1.6 to 1.3 is good, with 1 being the goal. The lower the number, the larger percentage of the total power furnished to the facility is used by the data center

operations. (80% of the power furnished used by the IT equipment would be very good. This is the DCIE indicator whose reciprocal is the PUE. 1/.8 = 1.25.)
Measure for Verification: Power Usage Effectiveness (PUE) = Total Building Power (kW)/IT Equipment Power (kW) and then Goal-PUE = $Y \times 100$ = Percent Reduction Needed to Achieve the Goal.

13.6 Developing an Electricity Baseline and Reducing Energy Consumption and Costs – A Case Study

The aim of this part is to illustrate to the reader how to conduct energy baseline for their facility and how to choose the equipment that must be part of energy centered maintenance model based on its energy classification process discussed in earlier sections.

The case study has been conducted on a multi-story commercial building, based on its electrical energy consumption. The facility has the following characteristics:

- Total build up area is 240,000.0 m^2.
- Total air-conditioned area is 220,000.0 m^2.
- Building use: commercial building.
- Building age: 4 years.
- Cooling source: district cooling (no chillers or cooling towers exist in the facility).

The common areas of the facility are served by multiple mechanical types of equipment as listed in the following table. The design connected load for each equipment in (kW) number of hours of operation each day (hr per day) is mentioned in table 13.22.

Table 13.22 Connected electrical load and operating hours.

List of equipment	Equipment code	Qty	Design power,kW	Operating hours, hrs/day
Primary chilled water pump	PCWP	4	95.0	16
Secondary chilled water pump	SCWP	12	34.0	16
Car park supply fan	CPSF	5	30.0	16
Car park extract fan	CPEF	5	30.0	16
Fan coil unit	FCU	363	2.0	16
Fresh air handling units 1	FAHU1	4	115	18
Toilet exhaust fan	TEF	4	6.0	18
Focus light	FL	24	1.0	6
Pole focus light	PFL1	64	0.15	6

List of equipment	Equipment code	Qty	Design power,kW	Operating hours, hrs/day
Pole focus light	PFL1	128	0.07	6
Elevator motor	EM	31	17.5	10
Escalator motor	ESM	2	15.0	12
Water booster pumps	WBP	6	7.5	4
Water transfer pumps	WTP	4	30	6
Water feature pumps	WFP	10	7.0	6
Irrigation pump	IP	2	10.0	4
Fresh air handling units 2	FAHU2	4	18.5	24
Fresh air handling units 3	FAHU3	5	31.0	24
Air handling units 2	AHU	4	10.0	18

13.7 Energy Baseline

It is required to gain the electrical energy consumption of the facility for the past three years for a more accurate baseline. The data shall be extracted from real-metering or utility energy bills.

Based on three years' historical data, the recorded electrical energy consumption is presented in table 13.23.

Table 13.23 Energy baseline.

	Electricity consumption, kWh			Average monthly ambient temperature, °C		
Year	**2013**	**2014**	**2015**	**2013**	**2014**	**2015**
Jan	800,076	744,194	714,240	21	19	21
Feb	824,274	778,312	746,271	22	20	24
Mar	840,575	798,681	742,376	25	24	25
Apr	861,632	809,592	792,806	29	30	28
May	953,582	886,523	868,506	31	33	34
Jun	954,811	888,682	816,048	33	35	35
Jul	1,027,979	955,307	935,282	37	37	38
Aug	933,077	875,176	813,035	37	37	38
Sep	920,051	820,823	746,207	34	35	35
Oct	863,218	799,391	743,678	31	32	32
Nov	868,583	795,642	739,416	26	26	27
Dec	793,503	737,728	644,569	22	22	22
Total	**10,643,374**	**9,892,065**	**9,304,449**			

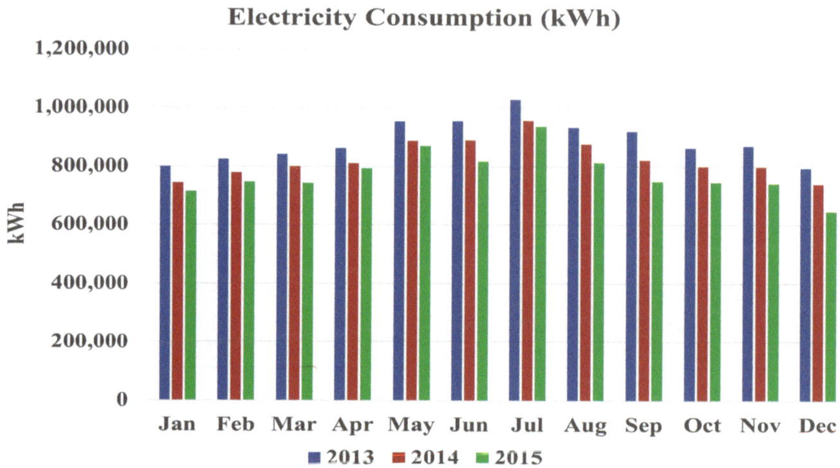

Figure 13.3 Electrical consumption trends by month.

A bar chart will show the trends between the years' kWhs by month. The chart will tell you if the years are similar and if they are seasonal. Figure 13.3 shows that 2015 has lower consumption and 2013 has the highest consumption. All four seasons – spring, summer, fall, and winter – are evident in the graph data.

All years are identical in trend, but the consumption is not the same. The year 2013 shows the highest consumption compared to other years, while the year 2015 shows the lowest consumption.

In late 2013, some energy reduction projects and energy centered maintenance model were implemented and resulted in reduction happening.

The year 2013 was considered as baseline for two years' energy reduction program, and the percent improvements in 2014 and 2015 when compared to 2013 are described in table 13.24.

Table 13.24 Percent improvement compared to 2013.

Total electricity consumption, kWh			Percent improvement compared to 2013		
2013	**2014**	**2015**	**2013**	**2014**	**2015**
10,643,374	9,892,065	9,304,449	Baseline	(6.7%)	(12.2%)

13.8 Energy Benchmarking

Energy benchmarking is an ongoing review of building energy consumption to determine if the building energy performance is improving or not.

In our case study, the benchmarking figures are developed by comparing 2013 energy utilization index (EUI) with the resulted values for years 2014 and 2015. This internal benchmarking compares the facilities' energy performance with itself in previous years, but it does not compare it to other similar facilities. Therefore, it is always recommended that the energy management team compares the facility energy performance with other similar buildings to decide if further improvements can be achieved.

Table 13.25 represents the EUI calculations. EUI is calculated based on total electricity consumed in the facility in a particular year, per total area of conditioned space.

The total air-conditioned area of the facility is 220,000.0 m^2.

Table 13.25 Energy utilization index for years 2013–2015.

Total electricity consumption, kWh			Energy utilization index ($kWh/m^2/year$)		
2013	**2014**	**2015**	**2013**	**2014**	**2015**
10,643,374	9,892,065	9,304,449	48.37	44.96	42.29

Based on the EUI mentioned in the above table and by comparing it to other similar facilities, the energy management team shall be capable of deciding if the performance of the facility is acceptable from energy consumption perspective or not. Accordingly, the team should decide if another energy reduction program should be implemented or not; if so, new energy reduction targets shall be set and a new energy audit should take place to define potential areas of savings.

If external benchmarking is used as a reference, the energy management team should carefully collect all information and data about the facility you are comparing with. There are many factors that can affect the energy consumption of the facility, such as operating hours, type of air-conditioning system, the size of air conditioning units, the age of the facility, maintenance strategy and maintenance quality, control logic of building automation system, glass to wall ratio, etc.

The EUI in 2015 was 42.29 $kWh/m^2/year$ and the energy utilization index for similar facilities is around 40.0 $kWh/m^2/year$. Based on that, it can be concluded that the facility is operating in a similar manner to other similar facilities. However, there are still potential areas of savings.

13.9 Energy Centered Maintenance Implementation

13.9.1 Step 1: Equipment Identification

Energy centered maintenance contributes to significant energy savings by ensuring that energy critical equipment are operating efficiently.

Table 13.26 Energy classification code.

	Energy classification	Energy impact	Description	Examples
Energy Classification Code	High 4 or 5	Large impact energy users	Systems with the following characteristics must be considered highly critical: • Operating profile: continuous running. • Energy capacity: high capacity.	Chilled water pumps, AHUs motors, lighting, etc. • Efficiency loss on these systems will result in high associated energy costs. • Must be continuously running.
	Medium 3	Medium impact energy users	Systems with the following characteristics must be considered critical: • Operating profile: continuous running. • Energy capacity: low capacity.	Toilet exhaust fans, lighting, etc. • Efficiency loss on these systems will have a medium impact on facility energy cost. • Might be running on time schedule.
	Low 1 and 2	Low impact energy users	Systems with the following characteristics must be considered non-critical: • Operating profile: non-continuous running. • Energy capacity: vary in capacity.	Fire pumps, stair case pressurization fans, emergency lighting, etc. • Might be high or low energy consumers. • Operates only in case of emergency.

The first step in this case study is to decide which equipment should be part of the energy centered maintenance analysis and which should be decided based on the given energy classification code (ECC) for each equipment. The equipment will be assigned with an ECC of 5 to 1 based on the following matrix and scale as shown in table 13.26 & figure 13.4.

The list of mechanical equipment serving the facility has been identified along with the connected electrical load and the total daily operating hours. Accordingly, the energy management team and maintenance personnel should be able to calculate the daily operating load in kW.hr.

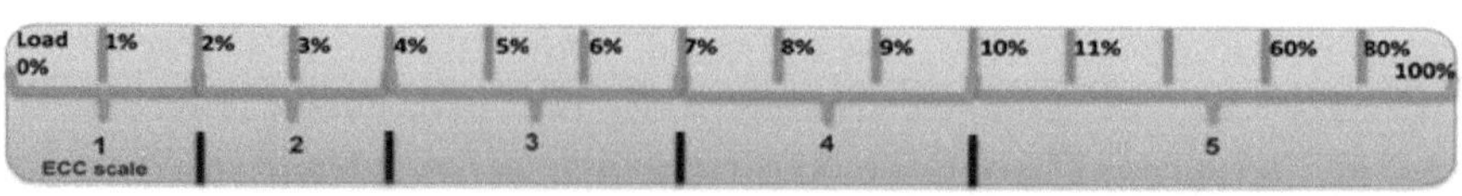

Figure 13.4 Energy classification code scale – Example 2.

Table 13.27 Total equipment's operating load.

List of equipment's	Equipment code	Qty	Design power, kW	Operating hours, hrs/day	Operating load for each equipment kW.hr	Operating load for all equipment kW.hr
Primary chilled water pump	PCWP	4	95.0	16	950	3800
Secondary chilled water pump	SCWP	12	34.0	16	340	4080
Car park supply air fan	CPSF	5	30.0	16	480	2400
Car park extract air fan	CPEF	5	30.0	16	480	2400
Fan coil unit	FCU	363	2.0	16	32	11,616
Fresh air handling units 1	FAHU1	4	115	18	2070	8280
Toilet exhaust fan	TEF	4	6.0	18	108	432
Focus light	FL	24	1.0	6	6	144
Pole focus light	PFL1	64	0.15	6	0.9	57.6
Pole focus light	PFL1	128	0.07	6	0.42	53.76
Elevator motor	EM	31	17.5	10	175	5425
Escalator motor	ESM	2	15.0	12	180	360
Water booster pumps	WBP	6	7.5	4	30	180
Water transfer pumps	WTP	4	30	6	180	720
Water feature pumps	WFP	10	7.0	6	42	420
Irrigation pump	IP	2	10.0	4	40	80
Fresh air handling units 2	FAHU2	4	18.5	24	444	1776
Fresh air handling units 3	FAHU3	5	31.0	24	744	3720
Air handling units 2	AHU	4	10.0	18	180	720
Total daily operating load for all type of equipment (kW.hr/day)						**51,392.36**

The total daily operating load of all types of equipment is 51,392.36 kW.hr/day. The operating load for each equipment is also mentioned in the same table 13.27 above. This information should be substituted in the following formula to decide the energy classification code for each equipment type.

$$\text{Single Equipment Load Percentage}(\%) = \frac{Quantity \times Operating\ Hours\left(\frac{Hours}{Day}\right) \times Connected\ Power(kW)}{Full\ Daily\ Load\ of\ All\ Equipment\left(\frac{kWh}{Day}\right)}$$

For example, for primary chilled water pumps, the load percentage should be:

$$\text{Single Equipment Load Percentage}(\%) = \frac{4 \times 10\left(\frac{hrs}{day}\right) \times 95\ (kW)}{51392.36\left(\frac{kWh}{Day}\right)} = 11.9\%$$

Accordingly, and by referring to the energy classification scale, the ECC for it is 5.

Considering the same calculations, the ECC for the rest of the equipment is shown in the table 13.28.

Table 13.28 Calculated energy classification code.

List of equipment's	Equipment code	Qty	Design power, kW	Operating hours, hrs/day	Load percentage %	ECC
Primary chilled water pump	PCWP	4	95.0	16	11.8%	5
Secondary chilled water pump	SCWP	12	34.0	16	12.7%	5
Car park supply fan	CPSF	5	30.0	16	4.7%	3
Car park extract fan	CPEF	5	30.0	16	4.7%	3
Fan coil unit	FCU	363	2.0	16	22.6%	5
Fresh air handling units 1	FAHU1	4	115	18	16.1%	5
Toilet exhaust fan	TEF	4	6.0	18	0.8%	1
Focus light	FL	24	1.0	6	0.3%	1
Pole focus light	PFL1	64	0.15	6	0.1%	1
Pole focus light	PFL1	128	0.07	6	0.1%	1
Elevator motor	EM	31	17.5	10	0.7%	1
Escalator motor	ESM	2	15.0	12	0.7%	1
Water booster pumps	WBP	6	7.5	4	0.4%	1
Water transfer pumps	WTP	4	30	6	1.4%	1
Water feature pumps	WFP	10	7.0	6	0.8%	1
Irrigation pump	IP	2	10.0	4	0.2%	1
Fresh air handling units 2	FAHU2	4	18.5	24	3.5%	2
Fresh air handling units 3	FAHU3	5	31.0	24	7.2%	4
Air handling units 2	AHU	4	10.0	18	1.4%	1

The ECM model focuses on all equipment with ECC of 5, 4, and 3 in its analysis; therefore, based on this case study, the following equipment will be selected for the analysis:

- Primary chilled water pump.
- Secondary chilled water pump
- Car park supply fan.
- Car park extract fan.

Fan coil unit.

- Fresh air handling units 2.
- Fresh air handling units 3.

In this case study, we have selected fresh air handling units 3 as an example to run the complete ECM model.

13.9.2 Step 2: Data Collection

The next step is to collect the operational parameters data about that equipment. Data can be collected from testing and commissioning figures, operation and maintenance manuals, asset registers, etc.

For fresh air handling units, the following data need to be gathered from testing and commissioning records:

- Supply/return fan airflow rate (m^3/hr).
- Motor power (Amps, Voltage).
- Cooling coil pressure drop (kPa).
- Cooling coil performance
- On-coil/off-coil temperatures (°C).
- Chilled water delta T (°C).
- Air on/off heat wheel temperature (°C).

Those data define the baseline for the measurements in the ECM model. The data collected are the expected operational parameters that the machine should deliver to meet its design intent.

For FAHU-3, the data are presented in table 13.29.

Table 13.29 Technical data for FAHU-3.

Section	Element	Data
Heat wheel Section	Supply air on heat wheel	28,000.0 m^3/hr
	Return air on heat wheel	18,000.0 m^3/hr
	air off-heat wheel temperature (dB/wB)	33.8/ 23.7 °C
	Motor connected load (kW)	0.20 kW
Supply fan section	Supply airflow rate	28,000.0 m^3/hr
	Motor connected power	18.5 kW
	Fan efficiency	73.0%

Table 13.29 Continued.

Section	Element	Data
Cooling coil section	Air on cooling coil temperature (dB/wB)	33.8/23.7 °C
	Air off cooling coil temperature (dB/wB)	13.8/13.0 °C
	Supply/return chilled water temperature	4.0/12.0 °C
	Water pressure drop inside cooling coil	35.0 kPa
Return fan section	Return airflow rate	18,000.0 m^3/hr
	Motor connected power	11.5 kW
	Fan efficiency	70.0%

13.9.3 Step 3: Identify ECM Inspections, Frequency, Craft, Tools, and Job Duration

After collecting the required data, the maintenance personnel should plan for ECM inspection and decide what performance parameters need to be measured.

Table 13.30 shows the inspections and measurements that should be conducted for FAHU-3:

Table 13.30 Inspection and measurements for FAHU-3.

Inspection	Tool	Craft	Duration
Measure airflow rate (m^3/hr)	Anemometer	Mechanical technician	20–30 minutes
Check motor's full load current (Amps)	Multi-meter	Electrician	10–20 minutes
Measure cooling coil performance on full load (air off-coil temperature °C)	Thermometer	Mechanical technician	10–15 minutes
Measure cooling coil performance (pressure drop, PSI)	Manometer	Mechanical technician	10–15 minutes
Measure two-way/three-way control valves response to space temperature	DDC simulator	Control technician	20–30 minutes
Measure chilled water temperature difference (delta T) °C	Thermometer	Mechanical technician	15–20 minutes

13.9.4 Step 4: Measuring Equipment Current Performance

The lists of data that have been collected in step 2 are the recorded performance parameters of the equipment according to testing and commissioning records. The maintenance personnel will now measure the current performance of the machine to compare the current readings to the first testing and commissioning readings.

Table 13.31 provides a sample data collection for FAHU-3 performance.

Table 13.31 Equipment current performance results.

Section	Element	Original reading (T&C)	Current reading (T&C)	Acceptable performance measures	Is the current reading acceptable?
Heat wheel section	Supply air on heat wheel	28,000.0 m³/hr	28,100.0 m³/hr	Current value = ±5% of testing and commissioning value	Yes
	Return air on heat wheel	18,000.0 m³/hr	18,100.0 m³/hr	Current value = ±5% of testing and commissioning value	Yes
	Air off-heat wheel temperature (dB/wB)	33.8/23.7 °C	34.1/23.9 °C	Current value = ±5% of testing and commissioning value	Yes
Supply fan section	Supply airflow rate	28,000.0 m³/hr	24,310.0 m³/hr	Current value = ±5% of testing and commissioning value	No
	Hourly motor operating load (kW.hr)	18.5 kW.hr	19.8 kW	Match value on data plate	No
Cooling coil section	Air on cooling coil temperature (dB/wB)	33.8/23.7 °C	33.8/23.7 °C	Current value = ± 5% of testing and commissioning value	Yes
	Air off cooling coil temperature (dB/wB)	13.8/13.0 °C	13.1/12.6 °C	Current value = ±5% of testing and commissioning value	Yes
	Supply/return chilled water temperature	4.0/12.0 °C	4.0/9.5 °C	100% ± 5% achieving desired delta T	No
	Water pressure drop inside cooling coil	35.0 kPa	34.5 kPa	Current value = ± 5% of testing and commissioning value	Yes
Return fan section	Return airflow rate	18,000.0 m³/hr	18,100.0 m³/hr	Current value = ±5% of testing and commissioning value	Yes
	Hourly motor operating load (kW.hr)	11.4 kW.hr	11.5 kW.hr	Match value on data plate	Yes

Based on the measurements taken, the following data are found to be inefficient:

- Supply airflow rate from supply fan.
- Supply fan motor power consumption.
- Difference between supply/return chilled water temperatures.

13.9.5 Step 5: Identifying Corrective Preventative Actions and Cost Effectiveness

The maintenance personnel are now aware of which parts of the FAHU are not performing according to design intent. Accordingly, the team shall conduct root-cause analysis to investigate the reasons of those deficiencies. Table 13.32 provides a sample for defining the problem found, cause, and corrective action taken.

Table 13.32 Problem, effect, root cause, and corrective action for FAHU-3.

Problem	Effect	Root cause	Corrective action	Is it cost-effective?
Supply airflow rate from supply fan, original flow rate is 28,000.0 m^3/hr, but current flow rate is 24,310.0 m^3/hr	Low cooling in served area	Increase in external static pressure due to closed motorized damper that is supposed to be normally open at one of the branches	Maintenance team checks all motorized dampers connected to this machine and damper open/close status was rectified	Calculated the associated total cost and found it to be cost-effective
Supply fan motor power consumption is expected to be 18.5 kW.hr, but the measured motor consumption is 19.8 kW.hr	High motor consumption and increased energy cost	Due to increase in external pressure, the motor consumed higher current to run the fan	Motor power consumption will be measured on a regular basis to ensure that it is not consuming higher energy	Calculated the associated total cost and found it to be cost-effective
Design chilled water delta T is 8.0 °C while measured delta T is 5.5 °C.	Low delta T penalties applied	Two-way valve actuator found stuck on opened position	Actuators status corrected and currently working on auto mode according to control logic	Minor cost lower than applied penalties

13.9.6 Step 6: Updating Preventative Maintenance Plans

When the corrective actions are implemented, it is recommended now to include those measures in the reliability preventive maintenance job plans, but those ECM tasks can be executed either annually or every six months based on the maintenance team analysis and equipment requirement.

For example, regular PPM plans have the following tasks:

- Cleaning FAHU filters.
- Tightening fan belt.
- Cleaning cooling coil.
- Lubricating shaft bearing.

And as a result of the ECM analysis, the following tasks should be implemented:

- Measure fan flow rate and confirm all dampers' functioning as required (annually).
- Measure motor power consumption and ensure it is optimum.
- Measure actual chilled water delta T and ensure it is matching design values.

14

Building Energy Centered Behavior Leading to an Energy Centered Culture

14.1 Kinds of Organizations' Cultures

Many consultants and authors of energy and environmental books and articles outline the need for employees' or students' engagement in energy consumption reduction, water consumption reduction, equipment failure prevention, solid waste reduction, and environmental stewardship. With the engagement, positive desired behaviors should result, and an appropriate culture will prevail. The cultures sought are an energy awareness culture, a water awareness culture, a recycling culture to save the planet, an equipment failure prevention culture, and an environmental stewardship culture. With sustainability being implemented in the main businesses, universities, and colleges today, there is a need for all of these cultures to be evident on site and practiced. These cultures together could be called a "sustainability culture" or an "energy centered culture."

14.2 Culture Definition and Building a Specific Culture

What is a culture? "It is the characteristics of a particular group of people, defined by everything from, religion, language, social behavior, manifestations, cuisine, music, and arts" (http://www.livescience.com/21478-what-is-culture-definition-of-culture.html). Characteristics, knowledge, language, and group of people apply to the energy centered culture we desire to have in place to our business, college, or organization.

Leadership: Management commitment is a must for success. Leaders are better when they have a vision or a policy.

Vision, Policy, and Goal:

- Vision: reaching a state of betterment.
- Policy: a commitment of managing what they are going to do or the employees are going to do.

Figure 14.1 Organizational culture.

- Goal: to improve something.

Values, Principles, and a Standard:

- Management desires for all personnel to possess and follow on a daily basis.
  Examples: committed to excellence, integrity, service before self, respect people, safety first, etc.

Attitudes and Morale: What managers and employees believe and express to others. The morale is determined by whether the personnel is positive or negative or somewhere in between.

Communications: Uses of several different media to inform, make aware, or teach organization's managers, employees, and contractors.

Roles, Responsibilities, and Measurement:

- Key players know what is respected of them and their authority. They have measures that show progress and results.

Business Processes: The way an organization gets its work done and achieves its mission and meets its customers' needs.

Training Awareness and Specialty Training: Create an awareness of the culture that is needed and why, plus explaining how each can help and support energy reduction. Develop skills in the organization's people that can help sustain energy management.

In other words, these eight attributes or characteristics are important to obtain the desired culture that remains in place as long as needed:

- Leadership.
- Vision, policy, and goals.
- Values, principles and standard.
- Attitudes and morale.
- Communications.

- Roles, responsibilities, and measurement.
- Business processes.
- Training.

All of the organizational cultural **characteristics that are shown in figure 14.1** can help achieve the desired culture, but they are not of equal importance. Leadership, vision, policy, goals, attitudes, morale, communications, roles, responsibilities, management, and training are essential to achieving the desired culture. These are interrelated as good leadership can impact favorably everyone's attitude and morale. Vision or Policy can help motivate the **group of people** to join and learn the **knowledge and language** and help accomplish actions that will contribute to goal attainment. Training is the method to depart the knowledge needed for everyone to perform specific desired tasks. Communications are the lifeblood that keeps the culture fluid, current on issues and tasks needed, plus knowing where current performance is and how it compares to the goals and vision. Roles, responsibilities, and management provide a structure; so everyone knows what their responsibilities are and what is expected of them.

Characteristics needed for a culture change or development:

- Leadership committed vision.
- Policy developed.
- Good positive attitudes.
- Roles and responsibilities defined.
- Both way communications.
- Training and commitment.
- Monitoring, recognizing, and rewarding.

Leadership with a vision or policy influences and changes attitudes, mindsets, and behaviors of people and helps them share a purpose and achieve operations and process improvements.

The goal is to transition the workplace into everyone unconsciously showing positive behaviors regarding their energy and water use, environmental stewardship, observing and reporting leaks, noise, and other items that stop energy waste. The possible phases or stages of change are to go from the state of being unconsciously and not participating to become cognizant of what to do (through training) and being consciously showing the desired behavior (driven by policies and procedures, being motivated, and being recognized and rewarded). Now the final stage of being unconscious but exhibiting positive and desired behavior automatically comes from the training to know what to do, their commitment, the practice of doing it over and over, and knowing they will be recognized and possibly rewarded for their participation.

The "What Behavior is Desired" comes through training. The main "What" in each of the cultured areas will be covered now. This list could be massive. Only four of the most common will be shown to get the point across how important it is that the participants know what is expected and what to do.

Water Consumption Reduction:

1. Report leaky faucets to maintenance or facilities:
   - Five drips a second is a steady stream.
   - One gallon of water = 15,140 drips.
   - A steady drip wastes one gallon every 15 minutes.
   - A steady trickle wastes one gallon every 4 minutes.
2. Reduce shower time to 5 minutes or less:
   - 150 gallons of water can be saved in a month by lowering the shower time by 1–2 minutes.
3. Stop water from continuously running while brushing your teeth or shaving:
   - Save up to 8 gallons of water a day by turning off the tap while you brush your teeth and shave.
4. Do not keep the water running while you are washing your auto:
   - A garden hose can use 15 liters of water per minute. That is 225 liters of water in just 15 minutes!

Energy Consumption Reduction:

1. Turn off the lights when you leave the room.
2. Use IT power management on your computer monitor and CPU:
   - Saves approximately $75 each in energy cost every year.
3. Unplug appliances when they are plugged in but not in use.
   - The appliances, electronics, and equipment use electricity when plugged in even though they are turned off.

Practice Energy Conservation:

1. Implement an energy conservation policy.
2. Develop an energy conservation program and implement it.

Solid Waste Reduction Culture:

1. Recycling paper, aluminum cans, plastics, and cardboard. Recycling 1 ton of papers saves 17 trees, 4000 kWh of electricity, 7000 gallons of water, over three cubic yards of landfill, and a lot of greenhouse gas emissions.

2. Recycle waste:
   - After computers, monitors, fax machines, printers, and electronics reach the end of life, they are recycled to save precious metals and keep them from going to the water table.

Environmental Stewardship:

1. Do not spill any chemicals:
   - Could get in water table if not cleaned up.
2. Do not pour chemicals or put old medicine down the sink or toilet:
   - Could do damage.
3. Participate as an ECO rep:
   - Help train others in environmental stewardship.
4. Serve on Green Teams:
   - Help reuse, recycle, and improve environmental stewardship.

Equipment Failure Prevention Culture:

1. Look for equipment running that may not be required to be running:
   - Motors often run 24/7 when maybe only needed for 8 hrs.
2. Listen for machinery making excessive noise:
   - Could need lubricating or a bearing replaced.
3. Look for lights with a lot of dust or other debris on them:
   - Clean and restore the lumens to what they should be.
4. Look for oil or water leaks:
   - Could be a severe environmental problem or loss of productivity due to an equipment failure if not caught and repaired.

Energy Centered Maintenance Culture:

1. Develop specific ECM model for your building.
2. Implement ECM model as part of your maintenance strategy.
3. Educate and train the maintenance personnel about ECM model.
4. Engage energy management team with maintenance team to optimize the energy consumption of the equipment.

15

Data Driven Energy Centered Maintenance Model

Data driven energy centered maintenance is the main component in developing a digitally enabled maintenance approach. That involves using software and hardware technologies for real-time monitoring of the equipment performance and comparing it to the historical performance trends that define a baseline of its ideal performance.

Implementing this model in buildings supports the organizations in their digital transformation strategy. It provides a business case for implementing cost-efficient maintenance tasks defined based on real-time data and real-time digital analytics.

Implementing this model allows the building operators to automate more than 50%–60% of the energy-related maintenance tasks, which increases the accuracy of predictive maintenance, reducing maintenance man-hours and expanding the equipment reliability, energy efficiency, and lifespan.

With the recent evolution of digital transformation in the maintenance industry and the availability of IoT devices and sensors connected to the building's equipment, big data analytics in the maintenance industry are becoming a key component in smart building operations.

15.1 Digital Transformation

One of the goals of energy centered maintenance is to identify and manage the equipment operation risks, which could run in a non-efficient manner and cause unnecessary expensive operational costs.

The current digital evolution trend that involves data driven decision-making has opened new opportunities for energy centered maintenance. Artificial intelligence and machine learning are helping the maintenance team to get to the next level of maintenance intelligence to provide early alarms related to the abnormal equipment performance that previously were not possible to be triggered.

Energy centered maintenance inspections can be conducted digitally by investing in IoT measurement sensors and instruments integrated with the building management system (BMS), migrated to the Cloud software, and gives real-time performance monitoring to the equipment. Those sensors will provide a significant amount of data that will be processed through data analytics software to facilitate the maintenance team's decision-making process and give them a direction about the best possible maintenance solution.

Automating energy centered maintenance tasks and inspections (such as measuring flow rates and power consumption of the equipment) provides real-time data about the equipment's energy performance. The readings give an actual assessment of the equipment performance. Accordingly, if an improvement is required, the building's operator can determine it. The data received on the building management system or the cloud can then be processed by machine learning and artificial intelligence software to help in the decision-making process.

Specially customized algorithms in AI and ML software recognize whether the sensors' values over a specific period fall outside the normal range. If such an event occurs, the system issues an alarm via the building management system to conduct corrective actions.

This kind of digital revolution in maintenance management has advantages in preserving long-term energy savings from the equipment consumption and improving the reliability of the equipment over time.

15.2 Digitally Enabled Energy Centered Maintenance Tasks

Data driven energy centered maintenance is the main component in developing a digitally enabled maintenance approach. That involves using software and hardware technologies for real-time monitoring of the equipment performance and comparing it to the historical performance trends that define a baseline of its ideal performance.

With an in-depth analysis of the regular ECM maintenance job plans, some of the maintenance inspections are automatable with additional sensors and devices, which will help monitor the components and enable implementing predictive maintenance strategy as part of energy centered maintenance model.

When combined with data analytical tools, data driven energy centered maintenance can predict future deterioration in asset performance and ultimately prescribe the most effectible corrective/preventative actions to prevent the asset from working in an inefficient matter.

Data driven ECM focuses on using data streaming to prioritize the maintenance resources and cost-effectively utilize them. This kind of approach will determine the equipment performance and act only when there is a change in the data trend.

The data driven method in energy centered maintenance pushes the data through statistical analysis or artificial intelligence and data analytics computing. Smart dashboards present the data that provides rich insight and best data visualization.

Predictive analytics software, such as machine learning and artificial intelligence, use algorithms to detect and analyze the equipment's operational performance. To predict any abnormality, the building's operators take corrective action ahead of time to prevent this abnormality from happening.

15.3 Benefits of Data Driven Energy Centered Maintenance

Data driven maintenance is not entirely new to the maintenance management industry. Still, it has become increasingly popular and virtually intrinsic to the significant megatrends in utilizing technologies for continuous performance monitoring and prediction. What is new about data driven ECM involves using real-time performance monitoring where instruments and measurements device continuously collects data about the asset. And the utilization of those data in a predictive matter (through AI and ML) allows for executing the right energy centered corrective maintenance actions that result in lower costs, more equipment reliability, and more energy-efficient operations.

With a data driven maintenance regime, abnormal equipment behavior that could soon lead to an increase in its energy consumption is identified independently of the usual planned preventive maintenance schedule. The building's operators rectify the problem before a significant impact on the equipment's performance, which reduces the equipment's operational costs.

The current digital evolution trend that involves data driven decision-making has opened new opportunities for energy centered maintenance model, which will help in achieving the following:

- Implementing the best maintenance strategy by utilizing the existing BMS system/software tools in the facilities.
- Using real-time analytics such as AI and ML for early fault detection, self-diagnostics, root-cause analysis, and deciding the best corrective action.

- Leveraging real-time operational parameters data for analytics.
- Increasing adaptation of predictive maintenance over planned preventive maintenance as part of energy centered maintenance model.
- Reducing maintenance and operation costs while increasing asset reliability and life span.
- Reduce human intervention in data analysis and reduce human error.
- Sustaining the equipment's energy efficiency.
- Improving and maintaining the operational performance of the machine.
- Providing the building operator with continuous data analysis helps in deciding immediate corrective actions in case of any performance deviation.
- Increase building occupant's satisfaction.
- Reducing maintenance man-hours and cost.

By permanently monitoring equipment's operational parameters using IoT sensors and devices, and after processing the collected data via the AI software, the equipment will be capable of conducting predictive analysis and makes decisions without human intervention.

15.4 Artificial Intelligence and Machine Learning in Energy Centered Maintenance

Artificial intelligence (AI) is the ability of a program or software to think and learn and conduct an analysis based on the data collected from sensors, predict abnormalities, and smartly decide based on it.

Machine learning (ML) is a type of algorithms and rules built-in software application to predict process results without being programmed. The main principle of machine learning is to develop algorithms that receive input data and use statistical analysis to predict behavior or output without human intervention. The machine then starts learning from itself and building more scenarios over time.

Machine-to-machine technologies are beginning to apply artificial intelligence, machine learning, and predictive analytics to make decisions on our behalf. All the data from these streams provide more significant insights, better equipment outcomes, new maintenance techniques, and more enormous operational cost savings.

The utilization of AI and ML in energy centered maintenance are capable of monitoring the equipment's current performance, collecting operational data, analyzing it, and identifying if there is any performance variation in its current condition compared to the required performance level. The system's intelligence focuses on data patterns to detect a potential change in the equipment performance based on similarities with predefined underperforming modes.

Vital data generated by the equipment pass to the analytics software (a cloud-based application or a building automation software) at periodic intervals, and analytics and machine learning software detect anomalies. Customized algorithms in these services recognize whether the sensors' values over a specific period fall outside the normal range. If such an event occurs, the system issues an alarm to the maintenance team.

An energy centered maintenance solution with machine learning will recognize the equipment's baseline performance and energy consumption levels and automatically set alert thresholds at the appropriate points. The AI and ML software will then provide recommended actions to be taken to restore the machine performance.

15.5 Model Capabilities

The data driven energy centered maintenance model algorithms enable the model for various capabilities that provide real-time data about the current operational condition of the equipment compared to its baseline operational parameters.

Building automation companies can develop the following models.

15.5.1 Operational Parameter's Deviation Model

Based on analytical model predictions, the model shall predict any deviation in the equipment operating parameters compared to the baseline. If the variation is about to start to happen, the model shall perform condition-based prediction and alarm generating to inform the building operators about this deviation.

15.5.2 Energy Loss Model

The prediction model will also identify the potential energy loss and increase in the equipment energy consumption due to the predicted deviation of the equipment's operational parameters compared to its baseline.

15.5.3 Self-Diagnostic and Root-Cause Analysis Model

The prediction model is based on defined rules that enable the problem's self-diagnostics and determines its root cause. Those rules will become more accurate as the machine learning model gets sufficient data to help more accurate predictions by time.

15.5.4 Remediation and Optimization Model

Based on the root cause found, the model shall be capable of generating tasks that should be performed by the building operator to put the equipment to its original condition and to optimize its operation and energy consumption.

The model will be capable of implementing an end-to-end monitoring solution for the building management system by collecting real-time data from the IoT sensors and devices and sending it to the cloud software to analyze it and to generate reports about the current operational parameter performance compared to its baseline.

15.6 Analytics Rules

Big data analytics is the backbone of predictive technologies. It is the science of analyzing raw data to reach conclusions about behaviors or patterns. In recent years, the role of these predictive technologies has become increasingly common in our daily lives. It potentially presents a considerable opportunity to improve equipment reliability and reduce maintenance costs in a building environment.

The analytics rules should be defined in a manner that can provide an automated process for the equipment's operational parameters deviation, which should be processed by the AI model built for that equipment to define the variation from its baseline and the reason behind this deviation.

The deviations in the equipment performance should be expressed in terms of mathematical analytics.

The analytical rules should define the deviation as a change in the equipment operational parameters, which could lead to an increase in its energy consumption. The deviation must be identified at its early occurrence or predicted in advance in order not to lead to a significant increase in the equipment energy consumption or reaching dissatisfactory conditions for building occupants.

The analytical rules should also consider the historical data available with the building operators in case data driven ECM model is developed for an existing building.

The data required for the analytical rules and model should be triggered by various IoT devices such as sensors and meters. They are either installed on the equipment or need to be installed. A list of these IoT devices is discussed in detail in the following section.

Customized algorithms recognize whether the values detected by the IoT sensors and devices in real time fall within or outside the normal operating range. If such an event occurs, the system issues an alarm to the building operators.

The data input from those IoT devices is provided on continuous bases to enable continuous condition monitoring and build standard behavior patterns. It will then be used as a source of the pool of data used in the analytical model rules to predict and identify any deviation from the standard behavior patterns.

15.7 Building Management System Schematics

This section represents the maintenance tasks that can be automated and monitored by the BMS system. A BMS schematic diagram for each of the equipment is added by identifying which sensors and devices can be installed inside the equipment to replace the manual inspections identified earlier in this book.

All the additional sensors need to be connected to the direct digital controllers and be programmed, tested, and commissioned. The sensors' data will be communicated to the BMS system's frontend, and real-time data will be processed and analyzed by the artificial intelligence software. When an abnormal activity happens or a change in the usual trend of the equipment operation is recorded, the system will create alarms to identify the maintenance team. Accordingly, corrective action shall be implemented after a root-cause analysis takes place.

The dominant reason for automating energy centered maintenance tasks is the use of equipment's primary operational data and real-time machine that quantifies when and what kind of maintenance is needed to maintain, repair, or replace critical parts within the equipment.

Following this method, maintenance organizations can track equipment performance and maintenance tasks can be performed or any other data that should be used to plan and schedule tasks that would prevent it from losing its energy efficiency.

Sensors, measurement devices, and microprocessor-based instrumentation, such as temperature sensors, airflow meters, and a multi-meter, can be used to monitor the operating condition of the critical energy equipment. The information and data received from these devices provide the means to manage the maintenance effectively.

This approach also has a financial benefit for the organization, adopting energy centered maintenance in its maintenance management system. Table 15.1 shows that one of the maintenance inspections is to measure the AHU flow rate manually, which will take around 20–30 minutes to complete. By using an airflow meter, the flow rate reading can be continuously monitored on the BMS system and saves the time of the manual measurements.

In principle, automating some of the energy centered maintenance tasks or inspections on BMS reduces the maintenance planned time and the maintenance costs, thus making this maintenance model more cost-effective.

Table 15.1 ECM inspection plan for air handling units.

Equipment type: air handling units				
Preventive maintenance				
Inspection	**Frequency**	**Tool**	**Craft**	**Duration**
Measure airflow rate (m^3/hr)	Annual	Anemometer	Mechanical technician	20–30 minutes
Check motor's full load current (Amps)	Quarterly	Multi-meter	Electrician	10–20 minutes
Measure cooling coil performance on full load (air off-coil temperature °C)	Annual	Thermometer	Mechanical technician	10–15 minutes
Measure cooling coil performance (pressure drop, PSI)	Annual	Manometer	Mechanical technician	10–15 minutes
Measure variable frequency drive effectiveness	Annual	Multi-meter	Electrician	45–60 minutes
Measure two-way/ three-way control valves response to space temperature	Quarterly	DDC simulator	Control technician/ electrician	20–30 minutes
Measure chilled water temperature difference (delta T) °C	Quarterly	Thermometer	Mechanical technician	15–20 minutes
Predictive maintenance				
Inspection	**Frequency**	**Tool**	**Craft**	**Duration**
Measure airflow rate (m^3/hr)	Continues	Flowmeter (connected on BMS)	BMS operator	5–10 minutes
Measure air off-coil temperature (°C)	Continues	Thermometer (connected on BMS)	BMS operator	5–10 minutes
Measure fan's motor power consumption (kWh)	Continues	Electricity meter	BMS operator	5–10 minutes
Measure chilled water temperature difference (delta T) °C	Continues	Thermometer	Mechanical technician	5–10 minutes

Traditional ECM tasks involve arriving at site, checking the system operation manually, taking measurements from the maintenance team, and then starting the essential maintenance works, which is a time-consuming process.

In this section, we identified which tasks can be done automatically by using the existing BMS to monitor and record specific reading for the following critical energy equipment:

- Air handling units.
- Fan coil units.
- Energy recovery units.
- Boilers.
- Pumps.
- Close control units.
- Fans.
- Cooling towers.
- Air cooled chillers.
- Chillers.
- Heat exchangers.
- Elevators.
- Motor control center – LV panel.

The introduction of automated maintenance tasks and data collection systems will ensure a continuous functional test of the equipment throughout its regular operation times to identify any energy efficiency issues as they occur. It allows the maintenance team to identify the root cause of the problems and resolve it in a shorter time, thus reducing maintenance costs and the equipment's energy consumption.

A wide range of sensors needs to be installed in the mechanical and electrical equipment for collecting data required for diagnosing the equipment condition and determining any abnormality in its operating parameters.

The schematic representation of the units is shown in the below figures.

The following list of sensors and instruments can be utilized to automate some of the ECM maintenance inspections:

- kWh meter.
- Airflow station.
- Water flowmeter.
- Temperature sensors.
- Pressure sensors.

Further, the tables below represent the job plan tasks, segregated into two categories – preventive maintenance tasks and predictive maintenance tasks. The tasks executed manually to retain the component condition are classified as the preventive maintenance tasks associated with a time-bound to conduct it. On the other side, the tasks are categorized under predictive maintenance tasks as the ones that can be automated and enable continuous monitoring of the equipment. Hence, if the maintenance tasks can be automated and are no longer required, the same will be removed from the regular maintenance plan and reduced maintenance time. At the end of each table, a time-saving figure is inserted, reflecting the potential cost savings.

And the BMS charts in figures 15.1 to 15.12 following tables 15.2–15.13 respectively, are showing how the equipment should be connected to the existing building automation system's controllers. The data collected by the sensors and the measurement instruments are being pushed to the analytics platforms continuously. This data is being analyzed by ML and AI software, where the intelligent algorithms are running to determine the healthy performance of the machines and to predict any abnormalities in the system.

Table 15.2 Air handling units.

Equipment type: air handling units			
Inspection	**PPM duration**	**PDM duration**	**Sensor type**
Measure airflow rate (m^3/hr)	20–30 minutes	Instant	Airflow station
Check motor's full load current (Amps)	10–20 minutes	10–20 minutes	
Measure cooling coil performance on full load (air off-coil temperature °C)	10–15 minutes	Instant	Temperature sensor
Measure cooling coil performance (pressure drop, PSI)	10–15 minutes	10–15 minutes	
Measure variable frequency drive effectiveness	45–60 minutes	45–60 minutes	
Measure two-way/three-way control valves response to space temperature	20–30 minutes	20–30 minutes	
Measure chilled water temperature difference (delta T) °C	15–20 minutes	Instant	Temperature sensor
Measure fan's motor power consumption (kWh)	15–20 minutes	Instant	kWh meter
Total time	**145–210 minutes**	**85–125 minutes**	
Time saving	**45.0%–60.0%**		

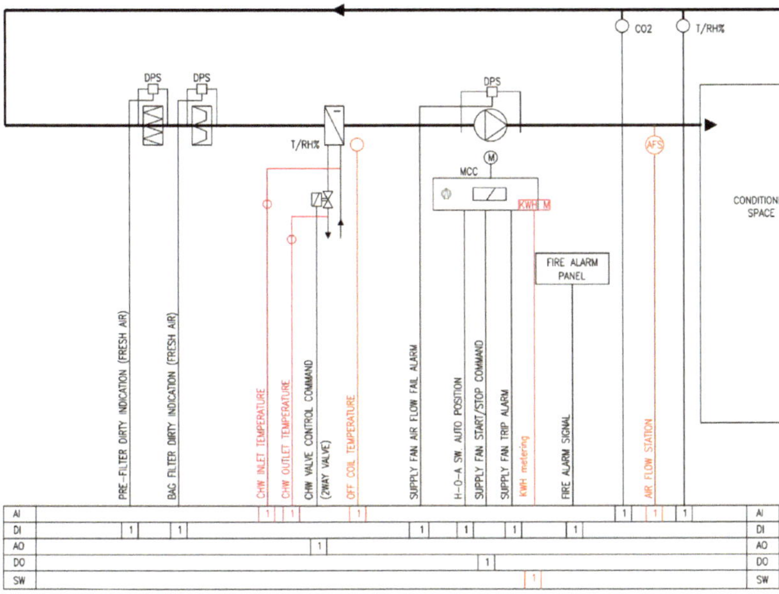

Figure 15.1 Air handling units – BMS schematics.

Table 15.3 Fan coil units.

Equipment type: fan coil unit			
Inspection	**PPM duration**	**PDM duration**	**Sensor type**
Measure airflow rate (m^3/hr)	20–30 minutes	Instant	Airflow station
Check motor's full load current (Amps)	10–20 minutes	10–20 minutes	
Measure cooling coil performance on full load (air off-coil temperature °C)	10–15 minutes	Instant	Temperature sensor
Measure cooling coil performance (pressure drop, PSI)	10–20 minutes	10–20 minutes	
Measure variable frequency drive effectiveness	10–20 minutes	10–20 minutes	
Measure two-way/three-way control valves response to space temperature	10–20 minutes	10–20 minutes	
Measure chilled water temperature difference (delta T) °C	15–20 minutes	Instant	Temperature sensor
Measure fan's motor power consumption (kWh)	15–20 minutes	Instant	kWh meter
Total time	**100–165 minutes**	**40–80 minutes**	
Time saving	**40.0%–49.0 %**		

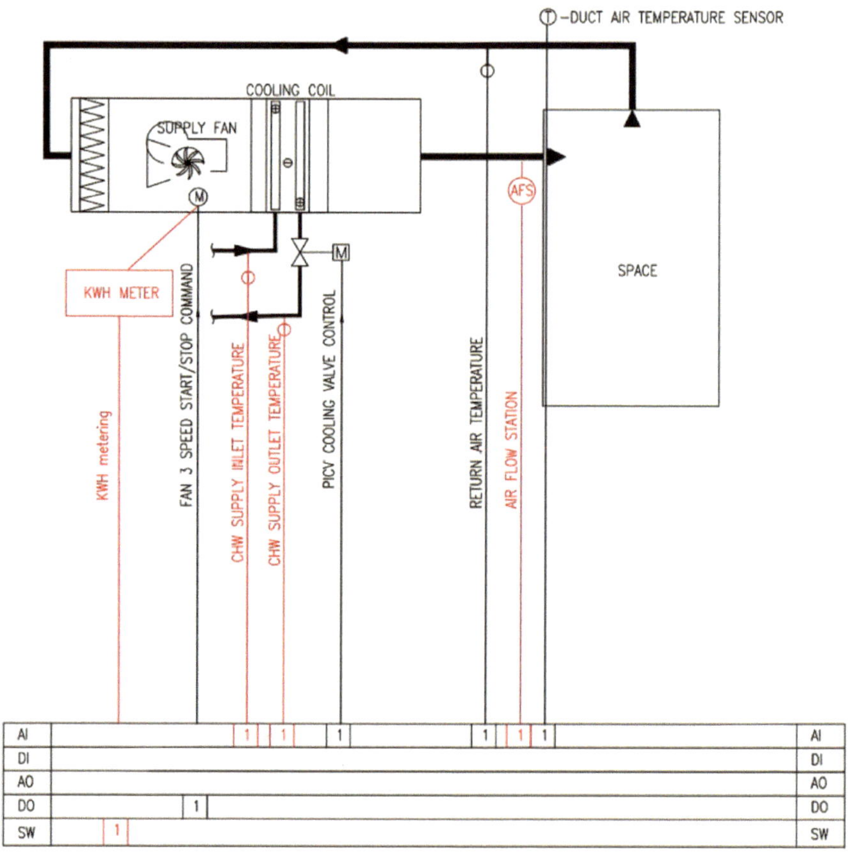

Figure 15.2 Fan coil units – BMS schematics.

Table 15.4 Energy recovery units (i.e., heat wheels).

Equipment type: energy recovery units			
Inspection	**PPM duration**	**PDM duration**	**Sensor type**
Measure airflow rate (m³/hr)	20–30 minutes	Instant	Airflow station
Check motor's full load current (Amps)	10–20 minutes	10–20 minutes	
Measure energy recovery performance on full load (air off-coil temperature °C)	10–15 minutes	Instant	Temperature sensor
Measure cooling coil performance (pressure drop, PSI)	10–20 minutes	10–20 min	
Measure fan's motor power consumption (kWh)	10–15 minutes	Instant	kWh meter
Total time	**60–100 minutes**	**20–40 minutes**	
Time saving	**60.0%–66.0%**		

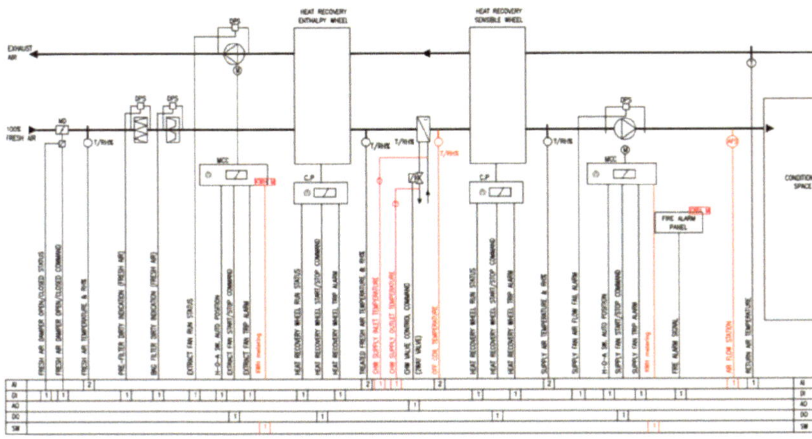

Figure 15.3 Energy recovery units – BMS schematics.

Table 15.5 Boilers.

Equipment type: boilers			
Inspection	**PPM duration**	**PDM duration**	**Sensor type**
Measure fuel combustion efficiency	30–45 minutes	30–45 minutes	
Inspect steam leakage	30–45 minutes	30–45 minutes	
Outlet water temperature (°C)	10–15 minutes	Instant	Temperature sensor
Primary system water pressure (PSI)	10–15 minutes	Instant	Pressure sensor
Total time	**80–120 minutes**	**60–90 minutes**	
Time saving	**25.0%**		

Table 15.6 Pumps.

Equipment type: chilled water pumps			
Inspection	**PPM duration**	**PDM duration**	**Sensor type**
Measure full speed water flow rate (gpm, lps)	60–90 minutes	Instant	Water flow station
Measure variable frequency drive effectiveness	45–60 minutes	45–60 minutes	
Measure pump's motor power consumption (kWh)	10–20 minutes	Instant	kWh meter
Total time	**115–170 minutes**	**45–60 minutes**	
Time saving	**60.0%–65.0%**		

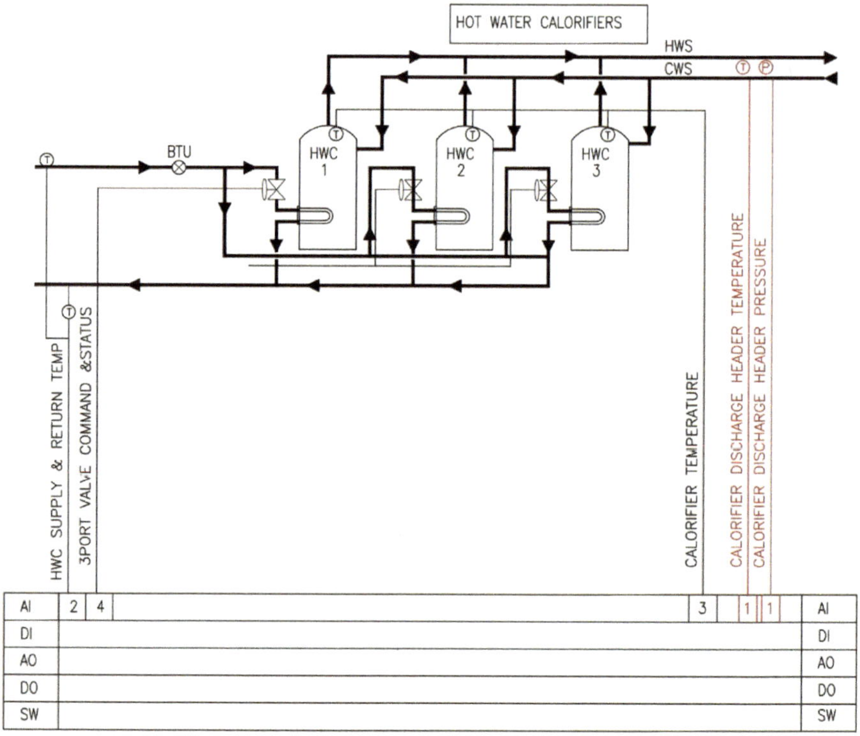

Figure 15.4 Boilers – BMS schematics.

Table 15.7 Close control units.

Equipment type: close control units			
Inspection	**PPM duration**	**PDM duration**	**Sensor type**
Measure airflow rate (m³/hr)	20–30 minutes	Instant	Airflow station
Check fan motor's full load current (Amps)	10–20 minutes	Instant	kWh meter
Check DX unit compressor full load current (Amps)	10–20 minutes	Instant	kWh meter
Measure cooling coil performance on full load (air off-coil temperature °C)	10–15 minutes	Instant	Temperature sensor
Measure cooling coil performance (pressure drop, PSI)	10–15 minutes	Instant	Pressure sensor
Measure variable frequency drive effectiveness	45–60 minutes	45–60 minutes	
Measure two-way/three-way control valves response to space temperature	20–30 minutes	20–30 minutes	

Equipment type: close control units			
Inspection	**PPM duration**	**PDM duration**	**Sensor type**
Measure chilled water temperature difference (delta T) °C	15–20 minutes	Instant	Temperature sensor
Measure DX unit energy efficiency ratio EER	30–45 minutes	30–45 minutes	
Total time	**170–255 minutes**	**95–125 minutes**	
Time saving	**44.0%–51.0%**		

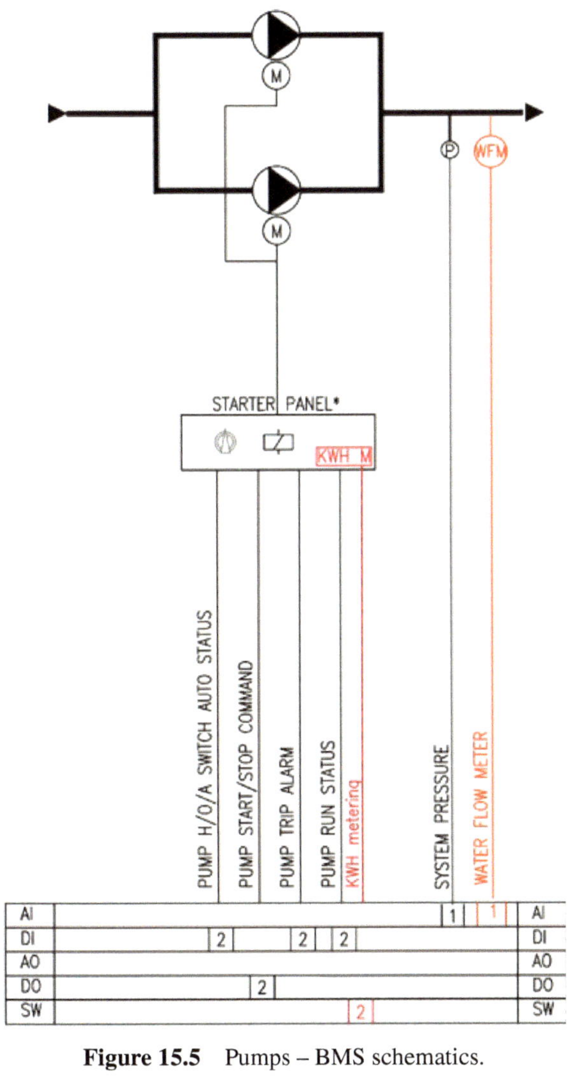

Figure 15.5 Pumps – BMS schematics.

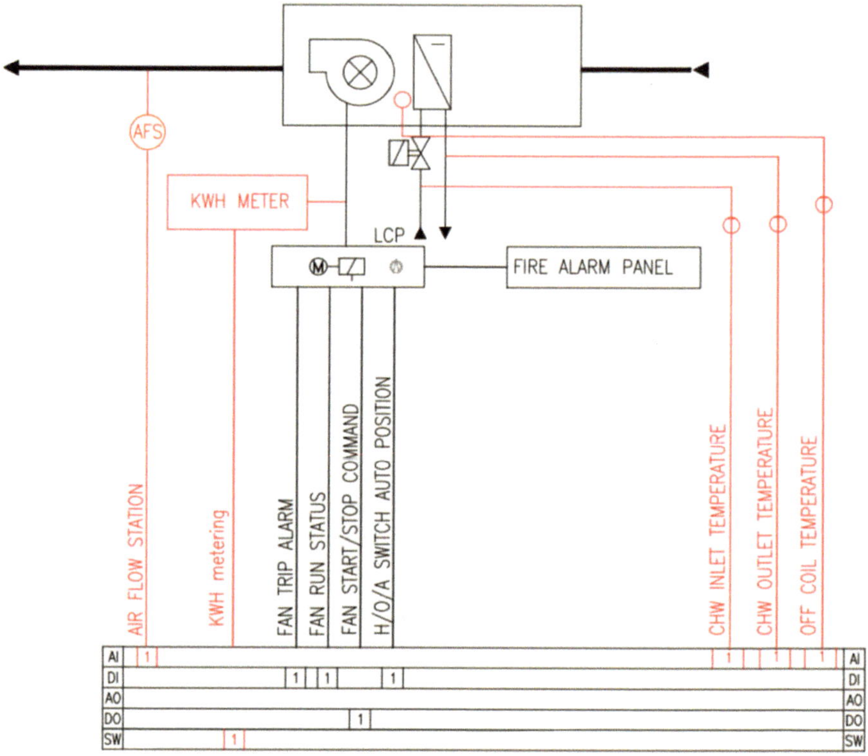

Figure 15.6 Closed control units – BMS schematics.

Table 15.8 Fans.

Equipment type: fans			
Inspection	**PPM duration**	**PDM duration**	**Sensor type**
Measure airflow rate (m³/hr)	20–30 minutes	Instant	Airflow station
Measure variable frequency drive effectiveness	45–60 minutes	45–60 minutes	
Measure pump's motor power consumption (kWh)	10–20 minutes	Instant	kWh meter
Total time	**75–110 minutes**	**45–60 minutes**	
Time saving	**40.0%–45.0%**		

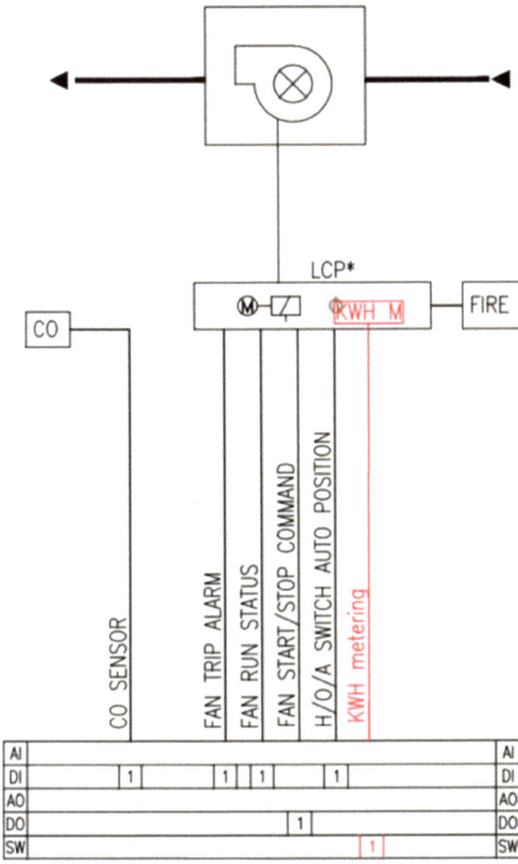

Figure 15.7 Fans – BMS schematics.

Table 15.9 Cooling towers.

Equipment type: cooling towers			
Inspection	**PPM duration**	**PDM duration**	**Sensor type**
Check fan motor's full load current (Amps)	20–30 minutes	Instant	kWh meter
Cooling tower range (in-out water temperature) – full load (°C)	10–20 minutes	Instant	Temperature sensor
Measure variable frequency drive effectiveness	45–60 minutes	45–60 minutes	
Total time	**75–110 minutes**	**45–60 minutes**	
Time saving	**40.0%–45.0%**		

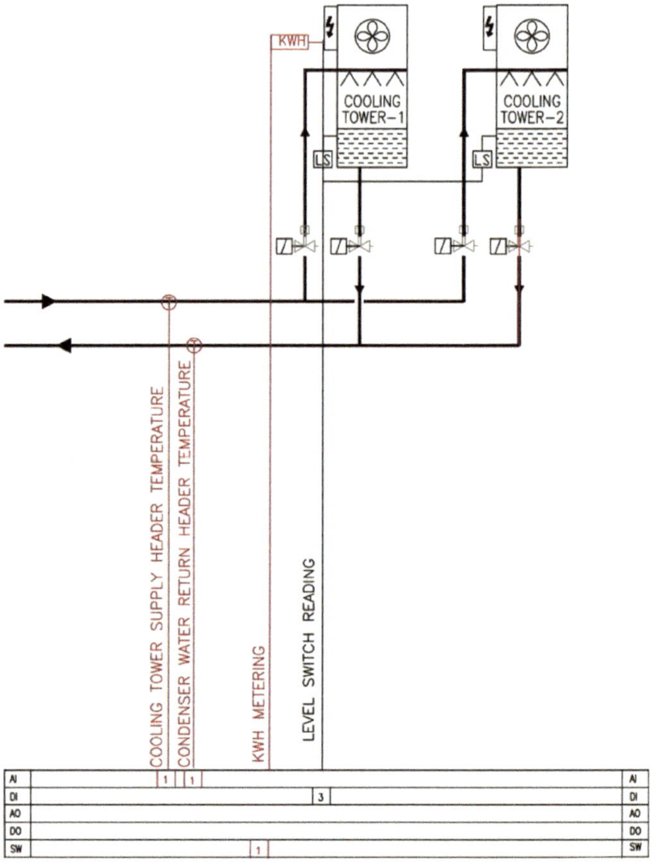

Figure 15.8 Cooling towers – BMS schematics.

Table 15.10 Chillers.

Equipment type: chillers			
Inspection	**PPM duration**	**PDM duration**	**Sensor type**
Check compressor motor's full load current (Amps)	10–20 minutes	Instant	kWh meter
Check condenser fan motor's full load current (Amps)	10–20 minutes	Instant	kWh meter
Evaporator pressure drop (PSI)	30–45 minutes	Instant	Pressure sensor
Refrigerant leaks test	30–45 minutes	Instant	Pressure sensor
Measure variable frequency drive effectiveness	45–60 minutes	45–60 minutes	
Total time	**125–190 minutes**	**45–60 minutes**	
Time saving	**64.0%–68.0%**		

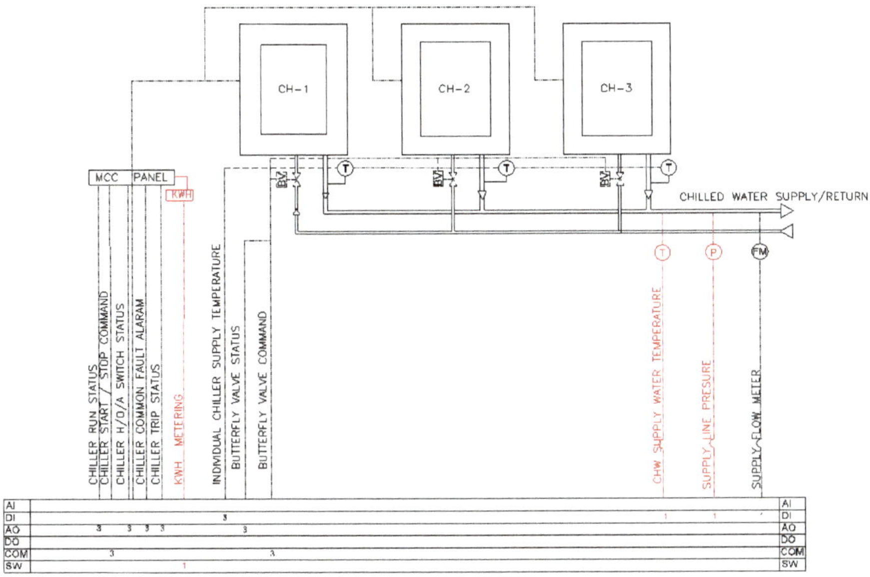

Figure 15.9 Chillers – BMS schematics.

Table 15.11 Heat exchangers.

Equipment type: heat exchangers			
Inspection	**PPM duration**	**PDM duration**	**Sensor type**
Measure pressure drop (PSI)	20–30 minutes	20–30 minutes	
Heat exchanger effectiveness (%) – (in/out temperature)	20–30 minutes	Instant	Temperature sensor
Total time	**40–60 minutes**	**20–30 minutes**	
Time saving	**50.0%**		

Table 15.12 Travelators and escalators.

Equipment type: travelators and escalators			
Inspection	**PPM duration**	**PDM duration**	**Sensor type**
Check motor's full load current (Amps)	20–30 minutes	Instant	kWh meter
Auto start/stop command	20–30 minutes	20–30 minutes	
Total time	**40–60 minutes**	**20–30 minutes**	
Time saving	**50.0%**		

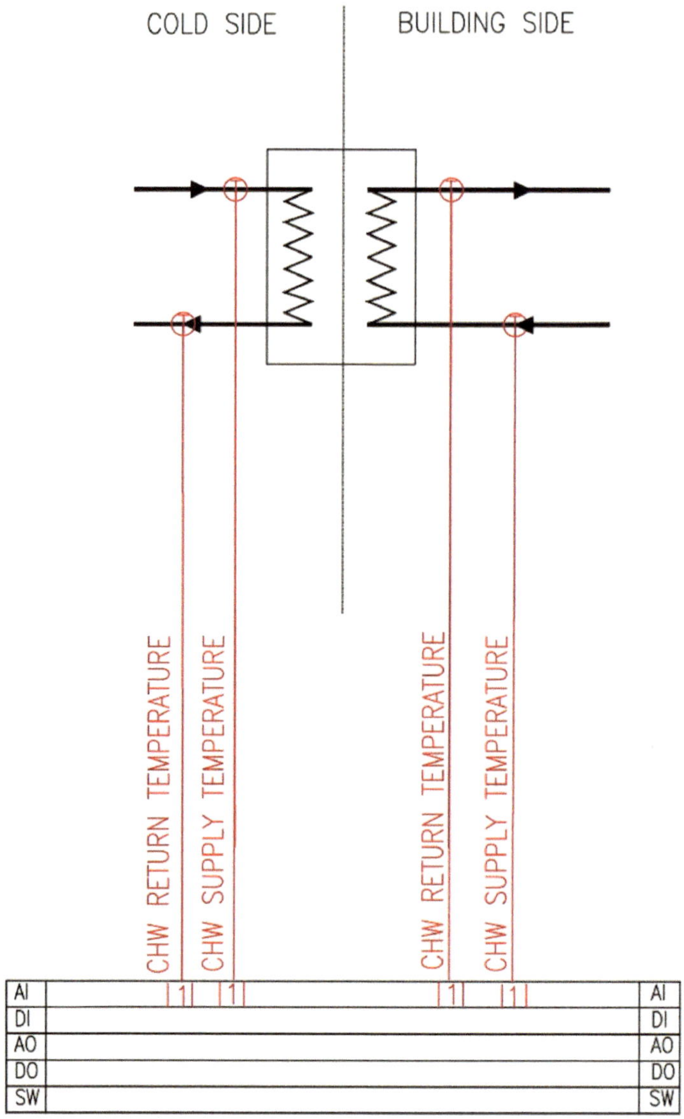

Figure 15.10 Heat exchangers – BMS schematics.

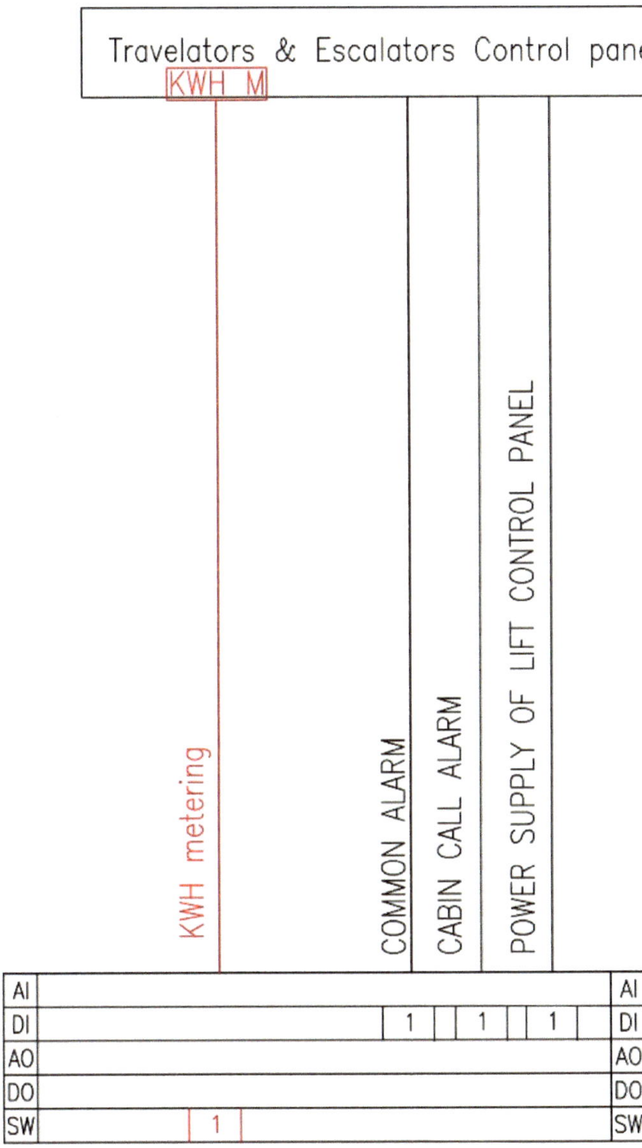

Figure 15.11 Travelators and escalators – BMS schematics.

Table 15.13 Motor control center – LV panel.

Equipment type: MCC – LV panel			
Inspection	**PPM duration**	**PDM duration**	**Sensor type**
In/out voltage at constant frequency	10–15 minutes	Instant	Voltammeter
Operating temperature (°C)	10–15 minutes	Instant	Temperature sensor
Total time	**20–30 minutes**	**Automated**	
Time saving	**Up to 100%**		

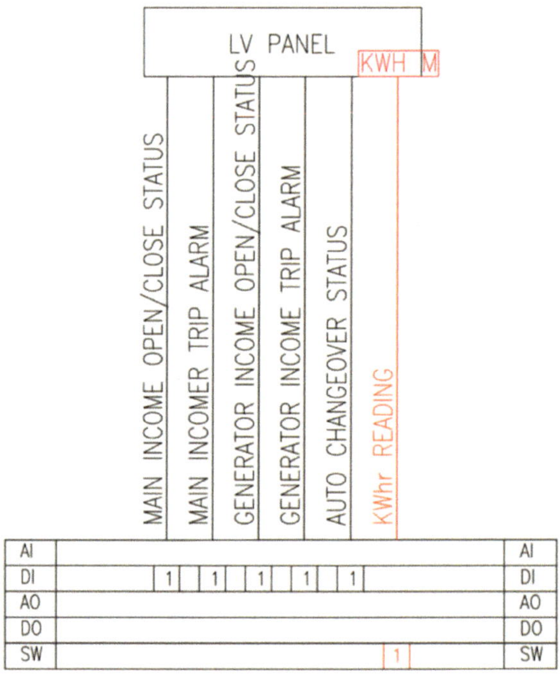

Figure 15.12 Motor control center – BMS schematics.

Data driven energy centered maintenance is a vital component of a comprehensive maintenance strategy that involves using analytical software technology for real-time monitoring of equipment operational parameters and comparing it to its baseline. Analytics software uses defined algorithms to detect operational performance variation via IoT sensors and devices connected to the building BMS as shown in the section.

Using advanced analytics and diagnostic technology and running it via artificial intelligence and machine learning software provides a tool that monitors critical energy assets to predict, diagnose, and recommend corrective actions continuously and in real time.

16

Conclusion

16.1 Designing and Implementing ECM

ECM provides the basis for identifying multiple low or no cost operation and maintenance practices that reduce energy consumption and improves the operational efficiency of the equipment. ECM works on the concept of returning the equipment to its original operational parameters (as originally commissioned) which result in improving its energy efficiency and reduce its energy use.

Energy centered maintenance model is a real maintenance program that focuses on energy-related equipment such as air handling units, electrical motors, pumps, etc. That equipment should be identified based on their energy consumption, and information should be collected from the testing and commissioning data (T&C) to compare the equipment's current behavior with its original commissioning data.

The proper execution of energy centered maintenance model should be delivered via planned job plans that are defined by the equipment's preventive and predictive maintenance needs. Equipment efficiency is significantly increased when tasks are accomplished by the standard proactive plans. Job plans should be designed for each piece of equipment based on relevant factors such as the maintenance tasks, frequency, tools, duration, and craft type. The cycle of designing and implementing the job plans is the main step in an energy centered maintenance process.

The ECM process's objective is to increase the energy efficiency of the equipment in a cost-effective manner using proven maintenance assessments and identifying maintenance related tasks that measure and improve the current operational behavior of the equipment.

Unlike reliability maintenance, the energy centered maintenance model does not intend to enhance equipment reliability or to prevent failures; the aim is to create maintenance tasks that prevent energy waste during equipment operation and to ensure that it is delivering the intended function.

Development and implementation of ECM model also have the following objectives:

- To provide educational practice of how energy consumption is related to maintenance.
- Improving maintenance regime to focus on the operational condition of the equipment.
- To identify any change in the equipment performance compared to testing and commissioning data.
- To identify improvements which can be made to increase equipment's efficiency.
- Optimizing the energy consumption of the equipment.
- Increase energy efficiency of the equipment through a low operation and maintenance cost.
- Ensure that equipment delivers the expected operational parameters as per the design intent.
- Reduce energy consumption of the facility.
- Reducing greenhouse gas emissions, mainly CO_2 emission and reducing carbon footprint, caused by energy consumption.

Years ago, after giving a speech on maintenance at an Industrial Engineering Conference, Marv Howell was asked: "Is there any Connection with Maintenance and Saving Energy." I had not been asked this question before. My answer is yes. If we paint a room white, it will take fewer lumens to serve the room. If we change a filter in an air conditioner on time, the air conditioner will not have to work as hard, thus reducing energy consumption. That question in the last 15 years has often been asked. Organizations that provide facility and equipment maintenance would like to advertise or be able to tell their client that they save energy while providing their maintenance. They were hesitant until ECM came along. ECM shows a direct relationship between reducing energy consumption and performing maintenance tasks. They knew that:

- Poor maintenance of energy-using systems, including significant energy users, is a major cause of energy waste in the government and the private sectors.
- Energy losses from motors not turning off when they should, steam, water and air leaks, inoperable controls, and other losses from inadequate maintenance are large.
- Uses energy consumption excess or energy waste as the primary criterion for determining specific maintenance or repair needs.

- Lack of maintenance tasks in measuring the operational efficiency of the equipment such as motor power consumption and equipment effectiveness.

ECM uses energy consumption excess or energy waste as the primary criterion for determining specific maintenance or repair needs. Energy centered maintenance (ECM) originated in 2012 when Marv Howell kept finding motors running 24/7 when they were only required to run 7–8 hrs daily. Also, he observed switches stuck on equipment, sensors not working, building automation systems with operators not trained, data centers using servers that were energy hogs, and cold air mixing wrongly with the hot air on the way to the computer room air conditioner (CRAC). He discovered that a maintenance program is needed to address equipment using excessive energy. In 2015, Fadi Al Shakhshir was working on developing a maintenance procedure that addresses what kind of maintenance tasks should be conducted during planned maintenance job plans that focus on the energy consumption of the equipment and the related operational parameters. As a final result, this book was written.

There are seven recognized maintenance types counting energy centered maintenance. They are:

1. Breakdown or reactive maintenance (before 1950, manufacturing revolution).
2. Preventative maintenance (1951).
3. Periodic maintenance (1951).
4. Predictive maintenance (around 1951).
5. Total productive maintenance (1951 origin, the 1980s in the USA).
6. Reliability centered maintenance (1960s origin, 1978 became known).
7. Energy centered maintenance (2012).

Each has advantages and disadvantages. The only one whose primary thrust is to reduce excessive energy use is energy centered maintenance. It is usually used in conjunction with preventative maintenance and predictive maintenance. This book outlines the steps to implement ECM and the equipment that should be included to maximize your energy savings.

The equipment to be included are:

I. Mechanical systems:
   1. Heating, ventilation, and air conditioning system:
      - Air handling units.
      - Fan coil units.

- Energy recovery units.
- Boilers.
- Pumps.
- Close control units.
- Fans.
- Cooling towers.
- Air cooled chiller.
- Water cooled chillers.
- Heat exchanger.
- Direct expansion air conditioning units.

2. Water supply system:
   - Pumps.
   - Heat exchangers.
   - PRV stations.
   - Boilers.
3. Drainage system:
   - Sump pumps (sewage).
   - Sewage treatment plant.
4. Storm water management system:
   - Rainwater pumps.
5. Building transportation system:
   - Elevators.
   - Travelators.
   - Escalators.

II. Fire-fighting systems:
- Fire pumps.

III. Electrical systems:
- Motor control centers.
- Variable frequency drive (VFD).

IV. Building management system:
- two-way/three-way valve functionality.
- Differential pressure switch – DPS.
- Differential pressure transmitter – DPS.
- Airflow meters.
- Velocity meters.
- On-coil temperature and humidity sensors.
- Off-coil temperature and humidity sensors.
- Space temperature and humidity sensors.
- Thermostat functionality.
- Control logic for all equipment

Each facility shall identify all systems and equipment applicable to their building; the technical information about the equipment can be found in asset registers, equipment schedules, O&M manuals, as-built drawings, etc. This information will be used to define the baseline performance of the energy-related equipment and to design a maintenance checklist and plan as part of the preventive and predictive energy centered maintenance strategy.

The different equipment selected impacts four general areas, each important to any facility. They are comfort, productivity, health, and safety.

Comfort – HVAC (heating, ventilation, and air conditioning), elevators and escalators, economizers, and lights and electrical system.

Productivity – Building transportation system, building management system, air compressors, etc.

Health – Drainage systems, storm water management system, and water supply system.

Safety – Fire-fighting system, ventilation, etc.

There are definite steps in developing and implementing ECM.

Step 1 is to identify the equipment to be included in ECM. Energy criticality of the equipment is defined based on two factors, the type of building system (i.e. AHU), and the equipment classification code. The classification code is based on the amount of energy consumed by the equipment and its operation

This step gives us the equipment classification codes that identify whether the equipment is included or not in ECM.

Step 2 is data collection and equipment operational baseline. All equipment that will be included in ECM must go through a data collection process. Data collection is needed to define the baseline for the equipment's operational parameters. The required data focuses on the operational parameters of the equipment. The type of data required is different for each type of equipment. The necessary data is found in testing and commissioning records, O&M manuals, as-built drawings and other maintenance records. An example of the data needed for air handling units is the fan flow rate (m^3/hr.), motor power (Volts, Amps), VFD data, control valve control logic, etc.

Step 3 is to identify ECM inspections, frequency, craft, tools, and job duration. The frequency of conducting ECM inspections differs based on the type of equipment and kind of maintenance. The inspections need to be carried out by qualified staff that is capable of conducting the inspections, recording the data, and making sound judgments in relations to the equipment behavior.

Maintenance records should include:

- A coherent equipment repair history.
- A record of maintenance performed on equipment.
- Cost of maintenance.
- Cost of energy.
- Replacement information.
- Modification information.
- Spare parts replacement.
- Diagnostic monitoring data (if available in BMS).
- Condition assessment.
- Energy efficiency records.
- Retro-commissioning records.

Maintenance records can be used for activities such as energy efficiency analysis, energy centered maintenance inspections, preventive maintenance tasking, predictive maintenance tasking, frequency planning, and life cycle analysis.

Specifying energy centered maintenance inspections should be established based on clear targets; for example, in an air handling unit, the target is to ensure that the AHU is capable of delivering the required airflow rate as intended in the design stage. This target sets what maintenance inspections should be conducted.

The inspections should be prepared considering the following:

- Determine which deficiencies may have an impact on customer satisfaction to correct.
- Determine which deficiencies are the most cost beneficial to correct.
- Determine which deficiencies are the most performance efficient to fix.
- Determine which performance parameters are critical for equipment operation to correct.

Scheduling energy centered maintenance inspections should be performed in such a way that energy centered maintenance tasks are conducted in the proper sequence, efficiently, and within prescribed time limits.

The frequency of energy centered maintenance inspections could vary based on different parameters such as:

- Equipment operating life.
- Physical condition.
- Failure interval and failure rate.

The more frequent the maintenance inspections take place, the higher the cost, but the greater the chances of maintaining equipment efficiency.

Conversely, the less frequent the inspection, the less the cost, but the higher the chances of increased energy use and increased energy waste intervals which result in high corrective maintenance cost. A balance between energy centered maintenance inspection, frequency, cost, and equipment efficiency should be assessed while defining the optimum ECM frequency.

The energy centered maintenance strategy calls to perform ECM inspections as part of equipment's regular preventive maintenance job plans; it will be either based on the following:

- Annual basis:
  - Where ECM inspections will be part of the annual PPM plans.
- Semi-annual basis:
  - Where ECM inspection will be part of the semi-annual PPM plans.

Energy centered maintenance inspections and job plans should be performed by a team of appropriately qualified and experienced personnel to achieve safe and efficient maintenance operations.

Experienced maintenance personnel should meet the following criteria:

- Understand general facility systems and equipment layout.
- Comprehend the purpose and importance of the facility's systems and equipment.
- Understand the effect of ECM work on the facility's systems.
- Assimilate industrial safety, including hazards associated with specific systems and equipment.
- Understand job-specific work practices.
- Comprehend maintenance policies and procedures.
- Be familiar with the personal protective equipment.
- Be capable of evaluating the performance of the equipment.

Identifying tools and specialized equipment that are required to execute an efficient energy centered maintenance inspection is an essential process that should be planned during the development of ECM job plans and inspections. A controlled supply of the proper type, quality, and quantity of tools and special equipment serves to avoid delays in maintenance work activities and increase worker efficiency. Defining the right tools is essential to allow the maintenance personnel to measure the current performance of the equipment which determines if any energy waste is noticed and to identify if the equipment is under performing, overperforming, or performing as intended.

The maintenance team should have a documented program for the control and calibration of test equipment and tools that ensure the availability of calibrated specialized equipment and tools.

Step 4 is measuring equipment's current performance and compare to baseline.

The maintenance personnel should be capable of measuring the current performance of the equipment during energy centered maintenance inspection. Current performance defines the actual operational condition of the equipment which will be compared to the baseline performance as recorded in the testing and commissioning phase.

Measuring the equipment's current performance involves collecting and analyzing actual data about the operational parameters of the equipment. Data such as equipment's efficiency, power consumption, will help in determining if any part of the equipment is not delivering its intended function, which, in its turn, results in identifying what kind of corrective actions should be done to improve and restore the operational efficiency of the equipment.

The measurements shall be compared with the baseline value, and the maintenance personnel shall be capable of judging if the equipment is overperforming, underperforming, or performing as intended. For example, measuring motor running current may be acceptable if it is operating within the acceptable range compared to testing and commissioning values, or an AHU is delivering an acceptable range of airflow compared to original data.

Step 5 is identifying corrective/preventive action and cost effectiveness.

When the root cause of a particular equipment performance deficiency is known, the right corrective action and repair can be implemented. Energy centered maintenance model focuses on identifying any operational deficiencies or energy waste; therefore, the corrective action may include specific repairs or replacements. Therefore the cost effectiveness of the repair needs to be determined prior to conducting the corrective action.

Identifying corrective action will help in improving the overall balanced, proactive maintenance strategy in the following:

- Identifying what needs to be done to restore equipment performance and what the expected results are.
- Identifying preventive actions to prevent the problem from being happening again.
- Identifying which maintenance process needs improvement.
- Identifying if new processes need to be implemented.
- Identifying new training for maintenance personnel.
- Identifying all costs associated with implementation and determining cost effectiveness.

Cost effectiveness should be calculated for each ECM tasks as well as corrective and preventive actions; calculating cost effectiveness should count for all maintenance activity costs associated with those actions, as well as potential energy reduction as defined in a certain period.

The maintenance costs include multiple elements such as:

- Time of maintenance.
- Labor cost.
- Material and consumable cost.
- Equipment cost.
- Calibration cost.
- Spare parts cost.

Energy saved could be calculated by different ways for each type of equipment by itself. For example, energy saved by enhancing motor efficiency equals the difference in motor's kWh before and after enhancement; this difference can then be converted to cost saving over a defined period (can be months, years, or the life cycle of the equipment) and can be compared to maintenance cost.

Restoring equipment performance is focusing on the following outputs:

- Maximizing equipment's operational efficiency.
- Restoring equipment's energy efficiency.
- Reducing energy waste by the equipment.
- Lowering operation cost by reducing energy consumption.
- Restoring the original performance of the equipment.
- Improving equipment quality.
- Ensuring that equipment is delivering its intended performance.
- Understanding the effect of equipment age, operating environment on its performance.
- Obtaining data for continuous improvement and high operational effectiveness.

Corrective/preventive actions are required to restore the performance of a failing equipment. It is necessary to verify that the corrective action was effective not only in eliminating the cause of the failure but also in restoring the equipment's performance.

Step 6 is updating PM plans on CMMS.

Once the energy centered maintenance approach has been established, regular use of the maintenance program should be implemented regarding scheduled preventive maintenance as well as predictive maintenance. The new maintenance program should be updated on the balanced proactive maintenance strategy to establish a proactive system that reviews the energy-efficient data of the equipment which helps to identify system operational performance before a major deficiency occurs.

The new maintenance program should be updated in the maintenance management system (CMMS) to ensure that all energy centered maintenance tasks that are defined during ECM inspection are now part of the preventive maintenance plans and predictive maintenance practice.

Updating PM plans requires the following:

- Identifying ECM tasks and frequency.
- Identifying and matching the appropriate skill sets to the tasks.
- Identifying the appropriate materials to the tasks.
- Identifying the appropriate tools/special equipment to the tasks.
- Identifying all other resources needed to perform the job.
- Uploading completed job plans to current CMMS.

The job plans provide all the details regarding safety, environmental and regulatory issues, as well as the operations, required downtime, affected components/systems, materials, labor, and tools required to do the work. The procedural part of the plan contains a task or a logical sequence of tasks, while each task consists of a list of steps.

In summary, the technical steps in developing and implementing ECM are:

Step 1. Identifying the equipment to be included in ECM.

Step 2. Data collection and equipment operational baseline.

Step 3. Identifying ECM inspections, frequency, craft, tools, and job duration.

Step 4. Measuring equipment's current performance and comparing to baseline.

Step 5. Identifying corrective/preventive action and cost effectiveness.

Step 6. Updating PM plans on CMMS.

The eight characteristics of a successful energy reduction program are shown next. Energy centered maintenance should be included in characteristic #6 as an energy efficiency measure.

16.2 Characteristics of a Successful Energy Reduction Program

Eight characteristics keep showing up at organizations that have been successful at reducing energy consumption and energy costs. They are:

1. Top management leadership supports, committed and involved in the energy reduction effort and becomes the programs GLUE (<u>**G**</u>ood <u>L</u>eaders <u>U</u>sing <u>E</u>nergy).

2. Energy reduction is made a corporate priority.
3. Corporate goals are established and communicated.
4. The energy champion or energy manager or both along with their cross-functional energy team select challenging strategies that include development of an energy plan and objectives and targets with actions plans.
5. Key performance indicators (KPIs) and key result indicators (KRIs) are employed and kept current and visible to measure and drive progress and results.
6. Sufficient resources are provided to fund or ensure adequate countermeasures are implemented to achieve the corporate goals.
7. An energy efficiency culture is achieved.
8. Sufficient reviews are conducted to ensure that continuous improvement, compliance to legal requirements, and adequate communications are provided to keep all stakeholders informed, motivated, and engaged.

16.3 Data Driven Energy Centered Maintenance

Data driven energy centered maintenance is a primary component in digitally enabled maintenance approach that involves using software and hardware technologies for real-time monitoring of the equipment performance.

Big data analytics is the backbone of predictive technologies. It is the science of analyzing raw data to reach conclusions about behaviors or patterns. In recent years, the role of these predictive technologies has become increasingly common in our daily lives. In a building environment, it potentially presents a considerable opportunity to improve equipment reliability and reduce maintenance costs.

Coupled with IoT, machine-to-machine interfaces, the cloud, and increased server capacity, large volumes of information can be shared quickly. These large volumes of data can be stored, analyzed, and used for building data models. Data driven energy centered maintenance, when combined with data analytical tools, can predict future deterioration in asset performance and, ultimately, prescribe the most effective corrective/preventative actions to prevent the asset from working in an inefficient matter.

The use of data analytics can also help deploy a meaningful forecast to fix equipment failures and energy inefficiencies, which is achievable with the help of integrated systems within a facility. Decision-making also becomes more efficient and effective with data sets collected from a more diverse range of sources such as sensors, smart meters, and IT networks.

Facilities can also leverage big data for energy savings and to optimize building performance.

Data driven ECM is based on using data streaming to prioritize the maintenance resources and to utilize them cost-effectively. This kind of approach will determine the equipment performance and acts when there is a change in the data trend to act.

The utilization of artificial intelligence (AI) and machine learning (ML) can be implemented to monitor the data generated by the equipment, analyze it, and identify any variation in the equipment performance. The intelligence of the system focuses on data patterns to detect a potential change in the equipment performance based on similarities with predefined underperforming modes.

Vital data generated by the equipment is pushed to the analytics software (which can be a cloud-based application or a building automation software) at periodic intervals. Analytics and machine learning software is used to detect anomalies. Unique customized algorithms in these services recognized whether the values detected by the sensors over a specific period fall outside the normal range or could not be recorded. If such an event occurs, the system issues an alarm to the maintenance team.

An energy centered maintenance solution with machine learning will recognize the equipment's baseline performance and energy consumption levels and automatically set alert thresholds at the appropriate points. The AI and ML software will then provide recommended actions to be taken to restore the machine performance.

For those custodians monitoring the health of building assets and planning maintenance activities, predictive maintenance, and big data analytics provide incredibly powerful insights. By viewing assets in dashboards, studying the hierarchy of assets, visualizing crucial key performance indicators, determining asset health, receiving alerts triggered by preconfigured rules, and seeing building performance at a single click, they gain an unrivaled level of insight.

ECM References

A Guide to Energy Audit – U.S. Department of Energy 2011

BS EN ISO 50001:2011Energy Management Systems.

CIBSE Guide F - Energy efficiency in buildings 2012

Degrading Chilled Water Plant Delta-T: Causes and Mitigations, by Steven T. Taylor, 2002.

Effective Implementation of an ISO 50001 Energy Management System (EnMS), Marvin T. Howell, Quality Press,2014

Energy Centered Management-A Guide to Reducing Energy Consumption and Cost, Marvin T. Howell, Taylor and Francis Group, 2015

Energy Centered Maintenance Modules 1 & 2, Association of Energy Engineers, Online Energy Seminar, instructor Marvin T. Howell, 2016

Energy Indicators that Drive Performance, Module 1 , Association of Energy Engineers, On Line Energy Seminar, instructor Marvin T. Howell, 2016

Harvard Manage Mentor. (n.d.). Gathering Performance Data. Retrieved October 21, 2009 from http://ww3.harvardbusiness.org/corporate/demos/hmm10/performance_measurement/set_targets.html

HVCA, Standard Maintenance Specification for Mechanical Services in Buildings, SFG 20.

HVAC Chilled Water Distribution Schemes, Continuing Education and Development, Inc. A. Bhatia.

http://www.oee.com/calculating-oee.html

http://search.aol.com/aol/image?q=organizational+change+models+and+theories&v_t=webmail-searchbox

https://studyacer.com/question/changing-behavior-case-study-analysis-410367

http://www.livescience.com/21478-what-is-culture-definition-of-culture.html

http://www.psychologicalselfhelp.org/Chapter11.pdf

HVCA, Standard Maintenance Specification for Mechanical Services in Buildings, SFG 20

Implementing Energy Efficiency for Measurable Results Modules 1 & 2, Association of Energy Engineers, Online Energy Seminar, instructor Marvin T. Howell, 2016

Low Hanging Fruit and Chasing Too Many Rabbits, Modules 1 & 2, Association of Energy Engineers, Online Energy Seminar, instructor Marvin T. Howell, 2016

Operations and maintenance Guide: Release 3.0, Energy.gov Office of Energy Efficiency & Renewable Energy https://www1.eere.energy.gov/femp/pdfs/OM_5.pdf

PECI. 1999. Operations and Maintenance Assessments. Portland Energy Conservation, Inc. Published by U.S. Environmental Protection Agency and U.S. Department of Energy, Washington, D.C

Phillips, L., Gray, R., Malinovsky, A., Rosowsky, M. (April 2009). *The Assessment Report: Documenting Findings and Using Results to Drive Improvement.* Texas A&M University Retrieved 10/12/09 from http://assessment.tamu.edu/wkshp_pres/AssessReport_UsingResults.pdf

PMMI Project. (August 2005). Target Setting — A Practical Guide. Retrieved October 21, 2009 from http://www.idea.gov.uk/idk/core/page.do?pageId=845670

PMMI Project. (August 2005). Target Setting Checklist. Retrieved October 21, 2009 from http://www.idea.gov.uk/idk/core/page.do?pageId=845670

List of Acronyms

ARCWT	Actual Return Chilled Water Temperature
ASCWT	Actual Supply Chilled Water Temperature
DRCWT	Design Return Chilled Water Temperature
DSCWT	Design Supply Chilled Water Temperature
AHU	Air Handling Unit
AI	Artificial Intelligence
BAT	Best Available Technique
BAS	Building Automation System
BMS	Building Management System
CSFs	Critical Success Factors
CCU	Close Control Unit
CRAC	Computer Room Air Conditioner
CUSUM	Cumulative Sum
CMMS	Computerized Maintenance Management System
CMMIS	Computerized Maintenance Management Information System
DPT	Differential Pressure Transmitter
DPS	Differential Pressure Switch
ECMs	Energy Conservation Measures
ECC	Energy Classification Code
ECM	Energy Centered Maintenance
EMS	Environmental Management Systems
EnMS	Energy Management Systems
EnPIs	Energy Performance Indicators
EPEAT	Energy Performance Environmental Attributes Tool
ESPC	Energy Savings Performance Contracts
EUI	Energy Utilization Index
ECI	Energy Cost Index
EPI	Energy Productivity Index
FCU	Fan Coil Unit
HVAC	Heating, Ventilation, and Air Conditioning
ISO	International Standards Organization
IT	Information Technology

IoT	Internet of Things
kWh	Kilowatt Hours
KPIs	Key Performance Indicators
KRIs	Key Result Indicators
LDT	Low Delta T (Low Temperature Difference)
ML	Machine Learning
MTBF	Mean Time Between Failures
MTTR	Mean Time To Repair
MCC	Motor Control Center
NGT	Nominal Group Technique
OEE	Overall Equipment Effectiveness
O&M	Operation and Maintenance
OEE	Overall Equipment Efficiency
PAL	For meetings develop Purpose, an Agenda, and Limit the time per agenda topic
P-D-C-A	Plan, Do, Check, Act the Deming Wheel
PE	Professional Engineer
PF	Power Factor
PIs	Performance Indicators
PRV	Pressure Reducing Valve
PPM	Planned Preventive Maintenance
PM	Planned Maintenance
PUE	Power Usage Effectiveness
RCA	Root-Cause Analysis
QMS	Quality Management System
QVS	Quality Value System Corporation
RCM	Reliability Centered Maintenance
SMART	Specific, Measurable, Actionable, Relevant, and Time-Framed
SWOT	Strength, Weakness, Opportunities, Threats
TPM	Total Productive Maintenance
T&C	Testing and Commissioning
UESC	Utility Energy Services Contract
VFD	Variable Frequency Drive

Index

About the Authors

Fadi S. Alshakhshir works in Emaar Facilities Management, in Emaar Properties PJSC, Dubai, United Arab Emirates. He graduated from Jordanian University of Science and Technology with Mechanical Engineering Degree, and from Heriot Watt University with M.Sc. degree in energy. He also has multiple professional certifications such as LEED Green Associate, Certified Energy Manager, Reliability-Centered Maintenance Facilitator, Six Sigma Black-Belt, Strategic Thinking and Planning, and EFQM Internal Assessor.

He has over 16 years of practical experience in design, construction, operation, and management of different types of facilities such as hospitals, hotels, malls, residential properties, and iconic buildings. He is always thinking of more efficient ways of managing facilities and buildings. He is very passionate about finding ways to reduce energy consumption, the impact on the environment by developing effective procedures and policies in place and implementing them. He is actively working on developing ideas and use cases that can improve the ways the maintenance is implemented in buildings by focusing on latest technologies that proved to be successful and cost efficient.

Marvin T. Howell is a former Senior Energy and Solar Consultant with SkyPower Energy Corporation in Arlington, TX, USA. As a contracted senior environmental associate with Analytical Services Inc. for eight years, Mr. Howell implemented environmental management systems at eight U.S. DEA facilities, including labs, division offices, an intelligence center, and an air operations center. In this capacity, he was instrumental in planning and designing several energy management initiatives that resulted in significant energy cost saving improvements throughout the DEA facilities. Previously, Mr. Howell was manager of distribution planning and reliability for Florida Power and Light, and he also served as a Lt. Colonel in Air Force civil engineering, where he was involved in energy reduction efforts, reliability centered maintenance, project management, construction and maintenance management, and efficiency and productivity improvements. He holds a BS degree in mechanical engineering from Mississippi State University and a master's degree in industrial engineering from the University of Pittsburgh. In 2014, he authored the book, *Effective Implementation of ISO 50001 Energy Management System*, published by ASQ Quality Press. In 2015, Marv authored the books, *Energy Centered Management, A Guide to Reducing Energy Consumption and Cost*, published by the Fairmont Press, Inc. and *The Results Facilitator, Expert, Manager, Mentor* published by CRC Press.